Electronic Telephone Projects

by

Anthony J. Caristi

Howard W. Sams & Co., Inc.
4300 WEST 62ND ST. INDIANAPOLIS, INDIANA 46268 USA

FIRST EDITION
FOURTH PRINTING—1982

International Standard Book Number: 0-672-21618-3
Library of Congress Catalog Card Number: 79-63868

Printed in the United States of America.

Preface

This handbook of electronic telephone projects was written to provide information long needed on the subject which could be used by electronics hobbyists and others to increase the utility of their telephones. Up to now much of this information was either unavailable to the general public, or appeared in bits and pieces in various electronics publications. With solid-state technology advancing at an accelerating rate, the telephone system has evolved into a communicating medium which is a far cry from the instrument Alexander Graham Bell invented.

Many of the devices presented here are available to the consumer for purchase or lease from the telephone company or independent manufacturers. Since it is the nature of electronics hobbyists to "build their own," this handbook should provide many hours of enjoyment for those who like to construct circuits and save money in the process. There are many of us who would not purchase a ready made unit, but will build a useful device at low cost. There is a certain satisfaction in seeing your homemade project come to life.

It is not necessary for the reader to have a formal education in electronics to construct and use the devices presented here. The text has been written with this in mind. For the technician or advanced hobbyist each project includes a concise summary of the theory of operation of the circuit. It is hoped that this handbook will provide the means for the reader to increase the usefulness of his telephone, whether his interest in electronics be casual or professional.

The author wishes to express thanks to Michael Sidey, Pat Sidari, and

Daniel Schacher for technical assistance in the preparation of this book. Many thanks also go to Karen Raia who cheerfully typed the entire manuscript.

Anthony J. Caristi

To my wife, Betty

Contents

SECTION 2—ELECTRONIC TELEPHONE PROJECTS

CHAPTER 5

This unit can be used to replace the loud, harsh telephone bell with a pleasing soft tone for use in locations where the normal telephone bell is too loud or annoying.

CHAPTER 6

The Telephone Sentry will provide positive indication if there is an additional telephone connected to the line besides the one being used at the time.

CHAPTER 7

This unit will permit both sides of a telephone conversation to be heard over a nearby fm radio.

CHAPTER 8

Each time the telephone is in operation the conversation from both parties can be recorded on an ordinary cassette tape recorder.

CHAPTER 9

This unit provides "hands free" operation of the telephone, and allows anyone in the room to hear the conversation.

CHAPTER 10

With this circuit an ordinary rotary dial telephone can be converted to Touch Tone operation.

CHAPTER 11

This unit will produce dial pulses in response to a push-button keyboard, for use in those telephone systems where Touch Tone is not operational.

CHAPTER 12

The telephone bell can be extended to anywhere a pair of wires can be run. It is battery operated, making it usable anywhere.

Section 1

General Information

Chapter 1

Basic Telephone Principles

When Alexander Graham Bell filed his application for a telephone patent on February 14, 1876, the apparatus he invented was extremely simple. It consisted of an electromagnetic transmitter and electromagnetic receiver connected together with a battery, and produced sounds too feeble for general applications in a practical telephone system. Through further research by Bell and others, a working telephone was realized by the development of the variable-contact carbon transmitter. This improvement provided a substantial increase in the audio frequency currents fed to the receiver, resulting in a satisfactory volume level. To this day, this method is still used in the basic telephone system for voice communications.

Before the telephone could be made practical for use over many miles of telephone cable, some method had to be used to automatically control the volume of sound so that it would be neither too loud for telephones located near each other, nor too weak for telephones far apart. This problem was overcome by the use of a variable resistance element, called a varistor, which decreases in resistance as the voltage across it increases. Thus, a varistor placed across the receiver of a telephone would automatically divert a portion of the audio frequency current to keep the receiver volume reasonably constant regardless of the distance between telephones.

Further improvements were made by the development of duplex or hybrid circuits in which most of the electrical energy developed by the transmitter is directed to the distant station with a minimum entering the speaker's receiver. A small multiwinding audio transformer and balancing network, placed inside each telephone, provides the full duplex operation described over a two wire circuit.

These circuit improvements over Alexander Graham Bell's original telephone are part of the modern day telephone (Fig. 1-1). Shown in Fig. 1-2 is a simplified schematic diagram of the Western Electric type K500 telephone set.

Fig. 1-1. A typical K500 telephone.

Part of the telephone circuitry, including the hybrid transformer and balancing network, is encapsulated in a small assembly with a group of teminals on top (Fig. 1-3). This is shown in Fig. 1-2 enclosed in dotted lines. Varistors placed within the encapsulated assembly serve to reduce the sound level when the called telephone is nearby so that satisfactory sound level is maintained. Another component in the encapsulated section is a capacitor which is used to resonate the ringer coils of the telephone to the 20 hertz ringing frequency. The ringing capacitor also serves to provide dc blocking to the dc voltage which is ever present across the telephone line.

The hookswitch is composed of three sections. These are illustrated in Fig. 1-2 as S1, S2, and S3. When the telephone is not in use (on the hook), S1 and S2 are open, and S3 is closed. This disconnects the entire telephone circuit, with the exception of the ringer coils and capacitor, from the telephone line and shorts out the receiver. The telephone

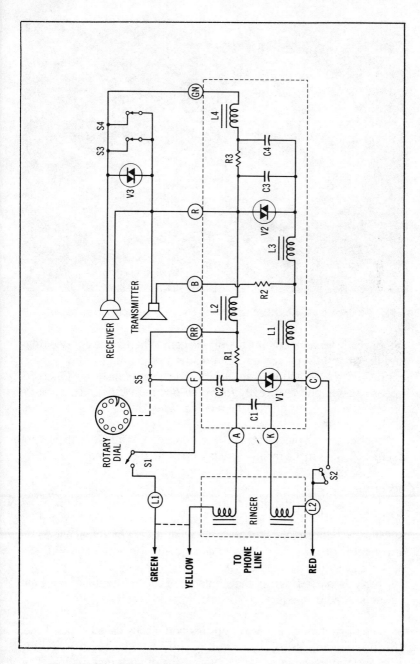

Fig. 1-2. Simplified schematic diagram of a type K500 telephone.

13

Fig. 1-3. Inside view of K500 telephone showing encapsulated assembly.

stands ready for an incoming telephone call. When the telephone rings and the handset is picked up, S3 remains closed until S1 and S2 have connected the telephone circuitry to the telephone line. Once S3 opens, the telephone is ready for voice communication. This feature prevents an annoying click in the receiver when the circuit is connected to the telephone line.

A second switch placed across the receiver, S4, is included to prevent a series of clicks in the receiver each time the dial is rotated. The dialing signal is actually a series of openings and closings of the circuit occurring at a rate of about ten times per second during the counterclockwise rotation of the dial. The number of openings occurring during dialing is determined by the number that is dialed, with the exception of the number zero. Dialing zero produces a series of ten openings in the circuit. When the dial returns to normal position, S4 opens and S5 remains closed. An RC circuit within the encapsulated assembly is placed across the contacts of S5 to attenuate any high frequency radio interference to nearby radio receivers.

A substantial improvement over the rotary telephone dial was accomplished through the use of push-button dialing called Touch Tone. This was made possible as a result of solid-state technology and the need for a more rapid and accurate means of customer dialing. This method also provides means for communication between the customer's telephone and other associated equipment such as computers. The rotary dial could not provide this service since it could control only the telephone company local central office equipment.

The push-button dial consists of a series of twelve push buttons, ten labeled 1 through 0, plus two additional buttons used for certain specialized functions. When any one button is pressed, a solid-state circuit inside the telephone set produces a pair of precise audio tones which are in the voice frequency range so that they can be transmitted wherever telephone communications are available. Fig. 1-4 illustrates the various pairs of frequencies generated for each of the twelve push buttons. The frequencies are divided into two bands, low band and high band, and each push button selects one frequency from each band. The system has been designed for future expansion to a total of sixteen push buttons. The four additional push buttons would control an eighth audio tone, 1633 hertz.

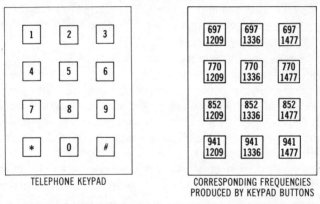

TELEPHONE KEYPAD

CORRESPONDING FREQUENCIES
PRODUCED BY KEYPAD BUTTONS

Fig. 1-4. Typical telephone touch-tone key pad and corresponding audio frequency tones generated.

The ringing signal is produced at the central office and is composed of a 90-volt rms 20-hertz signal which is impressed across the telephone line. This signal is produced by rotary generators called ringing machines, or by solid-state circuits which utilize no moving parts. The ringer inside the telephone set consists of a pair of coils wound around iron cores and a series capacitor connected before the hookswitch. The combination of inductance and capacitance of this series circuit is resonant at 20 hertz so that the ringer will operate with a minimum of ringing signal. Since the ringer circuit is connected across the telephone line at all times, the telephone company can determine if you have connected any unauthorized equipment across the line by simply measuring the impedance of the line. As a matter of fact, some telephone companies have an automatic computerized system which checks all lines on a daily basis, and alerts telephone company personnel if there is a change in the status of any line. This system is used to detect faults on telephone lines, as well as to detect unauthorized dvices. In order to legally connect an approved device to the telephone line you must notify the telephone company of the "ringer equivalence" of the device you intend to connect to the line.

The ringer circuit within the telephone can be connected to the telephone line in several ways, depending upon what type of service is being provided by the telephone company. For private lines, the ringer circuit is connected across the telephone line. This is accomplished by connecting the yellow and green wires together as shown in Fig. 1-2. When two party service is required, the telephone company provides selective ringing by feeding the ringing signal between one of the telephone lines and ground. In this way each party's ringer will be connected to a different wire, L1 or L2, and each will hear only the ringing signal belonging to him.

The two party ringing system can be expanded to a four party system with selective ringing by using a mechanical "bias" on the ringer armature and applying a combination of ac and dc ringing signal to L1 and L2 as required. Four party service was popular many years ago and may not be provided by some telephone companies any more.

Power to operate the telephone system, provided by the central office, is derived by a set of lead acid batteries connected in series to produce 48 volts dc. The battery is kept under constant trickle charge so that it will be able to carry the telephone system in the event of a power failure.

The telephone system has been designed so that it presents an impedance of about 1000 ohms between the dc source voltage at the central office and the customer's telephone set. When the telephone is "on the hook" the full battery voltage of 48 volts appears across the telephone line. When the telephone handset is lifted, the low impedance of the telephone set causes the voltage across the line to drop to about 6 volts. The dc current drawn by the telephone set is used as a signal to the central office that a call is to be initiated. This causes the telephone switching circuits to apply a dial tone to the line and await the dial pulses or tone frequencies. The low impedance of the telephone also serves to draw sufficient dc current from the line when a call is answered, so that the ringing signal is cancelled and the calling party is connected. The telephone company also provides an automatic delay feature which prevents a disconnection if the called party hangs up, accidentally or otherwise. This feature permits the called party to hang up one extension and pick up another. No disconnection will occur if the second telephone is picked up within 10 seconds. The calling party cannot take advantage of this feature, and will be disconnected immediately if the telephone is hung up.

Chapter 2

The Telephone Company and You

At one time, not too long ago, the telephone company had one position regarding customer-connected equipment being attached to the telephone line: It is not allowed. This one sided ruling meant that you could not legally obtain and connect a privately manufactured telephone and use it as an additional extension. Despite the telephone company's position in this matter, many people obtained telephones (Western Electric or otherwise), and connected them to the line for use as extensions. By doing this they avoided the additional monthly charges that were imposed upon telephone extensions. It was common knowledge that all one had to do to the illegal telephone was to disconnect the ringer circuit. This prevented the telephone company from detecting the additional extensions connected to the line, and although they knew that there were thousands of illegal telephones in use, they were powerless to do anything about it.

The situation remained pretty much the same until solid-state data communications and computer technology came into the scene. The invention of the transistor (ironically by Bell Laboratories), and later the integrated circuit, provided the means to design and manufacture new equipment which could be used in conjunction with the telephone network. One example of this is the time shared computer, which could be used by hundreds of customers located any distance away and all connected to the computer via the ubiquitous telephone lines. As new equipment for use with the telephone network became available, the telephone company reserved the right to be the only one who could legally lease it and connect it to the telephone system. This they did, and charged a monthly tariff for its use.

With more and more manufacturers entering the telephone accessory field pressure was exerted on the Federal Communications Commission to change the laws that gave the telephone companies a monopoly on telephone equipment. A milestone was reached in 1968 with FCC Tariff 263, commonly known as the Carterfone Decision, which permitted the connection of nontelephone company owned equipment to the telephone network. Since that time a proliferation of various telephone accessories has been designed and placed on the market for consumer use. Although the user was permitted to buy and own his telephone accessories, the telephone company required that a special coupling device be placed between the telephone network and accessory to prevent damage to the telephone network, or danger to its employees. Of course, a monthly fee was charged for use of the coupling device. The use of a coupling device for all consumer owned telephone equipment was an unfair requirement imposed by the telephone company. In some cases the company would buy the identical equipment and connect it to customers' lines with no protective device.

The FCC continued to work on the problem, and in their Docket No. 19528, a registration program was established to allow users of the nationwide telephone network to connect terminal equipment to the network without protective couplers. This requirement shifted the burden of protection of the telephone network to the manufacturer of the equipment. Before any such equipment could be used by consumers, the manufacturer had to submit to the FCC evidence that his equipment met standards that were established by the FCC. This registration program ensured that any telephone accessories connected to the telephone network would cause no damage nor harm any telephone company employee in the event of a malfunction of the equipment. One of the requirements of the registration program is the specification of "ringer equivalence number" which must be determined by the manufacturer of the equipment. Anyone who connects approved equipment to the telephone network must notify his telephone company that he is making the connection, and the ringer equivalence number for the equipment must be specified. This information allows the telephone company to determine the load presented to the telephone line when the ringing signal is impressed across it. Depending upon the type of equipment installed by the user, there may or may not be a monthly service charge imposed by the telephone company.

Prior to the enactment of the Carterfone Decision the do-it-yourselfer and independent manufacturers were limited to only two ways to legally connect nontelephone company owned equipment to the telephone system. Any such equipment had to be either inductively or acoustically connected. You may have seen early models of data communications systems in which the telephone handset was placed in a special cradle of the equipment. This was an example of acoustical coupling. The do-it-yourselfer, being limited in technical means, had to use only inductive coupling. Hard wiring to the telephone lines was illegal, although it probably was used more times than the telephone company would like to admit.

Today telephone accessory equipment may be connected to the telephone system in several ways. The telephone company may or may not use a coupling device, and they may permit you to make the connection yourself if you are already equipped with the telephone company installed receptacles or modular jacks. Depending upon what equipment you are using and what telephone company you are dealing with, there are three methods by which privately owned equipment can be legally connected to the telephone system.

The most elaborate method uses a STC coupler which is an active device and uses a telephone company provided power source. This

Fig. 2-1. Comparison of four-prong telephone plug and new modular plug being used by the telephone company.

type of coupling arrangement allows all normal telephone functions (dialing, ringing, talking, and receiving) to be used. A simpler type of coupling arrangement is the QKT coupler which is commonly known as the "phone patch." This coupler does not permit the use of the dialing and ringing functions of the equipment. Since some devices do not utilize such features the QKT coupler is perfectly adequate. The third method of connecting equipment to the line is the direct connection. This is the way most store bought equipment is connected to the telephone line. It is simply plugged into a telephone company provided receptacle. Before making direct connections, check with your telephone company to be sure it is permitted in your area.

The FCC also requires that any terminal equipment be connected to the telephone network through standard plugs and jacks. This decision was based on a long history of untrained telephone users installing equipment such as extension telephones in this manner without causing harm to the telephone network. It is a good rule and should be followed without exception. The old standard four-prong telephone plug and jack has been replaced by a modern miniature connector called a modular plug or jack (Fig. 2-1). Either of these types can be used and they are readily available from many electronics parts supply houses.

Chapter 3

Electronic Construction Techniques

The projects in this handbook make use of a printed circuit board on which to mount and connect most of the components. This section will discuss some methods of making your own printed circuits as well as other construction techniques. It is not difficult, and the necessary materials are readily available from electronics parts supply houses (Fig. 3-1). Even if you have never fabricated a printed circuit board before, with a little practice you will be able to produce printed circuits of professional quality.

Printed circuit boards can be constructed by several methods, depending upon the quantity of similar boards to be made and the techniques available to the builder. There are four basic methods which can be used. These are direct masking, photo mask/cut and peel, "lift-it" method, and professional artwork. The easiest method to use, especially when only one board is to be made, is the direct masking technique (Fig. 3-2). Emphasis will be placed on this method of construction. For general information, the following discussion is a brief outline of the other three methods of printed circuit construction. These three methods require sensitized copper-clad material.

The photo mask/cut and peel method can be used when you have a simple printed circuit layout which can be traced using a sharp knife. With this method it will be possible to construct as many boards as you like, using the photo mask produced by the cut and peel method and sensitized printed circuit board material. The basic film used for this

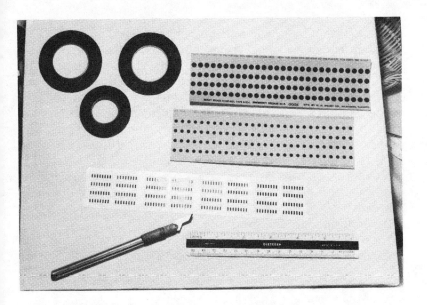

Fig. 3-1. Artwork materials and tools for printed circuit construction.

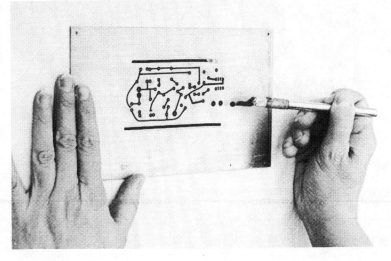

Fig. 3-2. Direct masking of printed circuit board (before etching.)

method is a thin, red adhesive backed plastic material laminated to a clear plastic base. With this method, a photo negative is produced by placing the plastic film over the printed circuit layout, tracing the copper foil conductors with a sharp knife, and lifting off the red film from the clear plastic base. Wherever the red film has been removed from the base the final printed circuit board will have a copper path.

The "lift-it" method of copying a published printed circuit layout consists of actually lifting the printed image from the paper and transferring it to a transparent film. This is accomplished by applying a special liquid to the printed circuit pattern in a series of six coats. This produces a transparent film which can be handled. The paper backing of the printed page is then removed by soaking it in warm water. This softens the paper so that it can be easily rubbed off the film, leaving the printed pattern. With this method a positive transparency is produced that can be used with specially sensitized printed circuit material designed for positive transparencies.

The professional artwork method of printed circuit construction (Fig. 3-3) is the method used by most manufacturers of printed circuit boards.

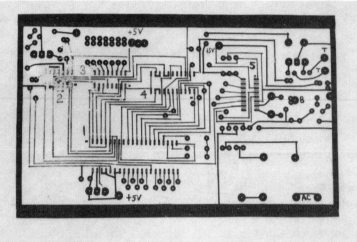

Fig. 3-3. Example of professional artwork method of printed circuit construction.

With this method, extremely accurate, high density printed circuit boards can be constructed since the artwork can be produced 2 or 4 times the size of the final printed circut board. This method consists of applying tapes, donut pads, multipad configurations (for integrated circuits), letters, numbers, and words to a sheet of mylar. This artwork is then photographed and reduced in size if necessary to produce a negative which can be used with sensitized printed circuit material.

The three methods of printed circuit construction just described all lend themselves to making more than one printed circuit board from the transparency produced. When it is desired to make only one printed circuit board, the direct masking technique is the fastest and easiest method to use. It requires no sensitized printed circuit material. For this method you can use ⅛ inch (3.2 millimeters) diameter donut pads and 0.040 (1.016 millimeters) inch wide flexible tape which is readily available from electronics supply houses. These dimensions are not critical, and in most cases you will be able to use other sizes of donut pads and tapes if that is all you have. Most of the printed circuit patterns shown in this handbook are full size, which will give you an idea of what size pads and tapes you will need.

A second, less desirable method can be used to lay out the printed circuit, and that is to use a special pen and ink, called resist, to draw the lines and pads directly on the copper. This method will produce a perfectly good printed circuit, but its appearance will not be professional since it is not possible to draw smooth lines. The big advantage of this method is that it requires fewer materials and is faster than using donut pads and tapes.

Copper-clad printed circuit material comes in several grades and thicknesses. The best choice would be 1/16 (1.59 millimeters) inch thick glass epoxy board material coated with one ounce (28.35 grams) copper on one side. One ounce (28.35 grams) copper refers to the weight of the copper in a one square foot (.00929 square meter) area and is about 0.0015 inch (0.0381 millimeter) thick. You can also use the lower quality phenolic material if that is all that is available to you, but glass epoxy would be a better choice since the cost of the board material is relatively low even for the best grade. You can also use printed circuit material which is copper clad on both sides, even though all printed circuits in this book are designed for single-sided boards. The etching time for double-sided material will be greater than that for single-sided material since there will be about twice as much copper that will have to be removed from the board during the etching process.

The chemical used for etching is a solution of ferric chloride. This etchant is available in a premixed state to the right strength for printed circuit work. Only a small amount, approximately 8 ounces (237cc), is required to etch one printed circuit board, and the remainder may be returned to the container for future use.

Before laying out the circuit on the copper-clad material take a piece of fine steel wool and polish the copper so that it is shiny and bright. This will remove any oxidation or dirt from the surface of the board and will permit more uniform and faster etching.

If possible, make a photocopy of the printed circuit layout and schematic diagrm of the project you are constructing. This will permit you to work directly from the copy, and not from the book. It is easier and avoids wear and tear on the book which you will want to keep for future reference. You can lay the printed circuit photocopy directly over the printed circuit material and locate all holes by pressing a sharp instrument directly through the photocopy. Once the holes are located in this manner it will be very simple to apply the donut pads and connect them together with the tape.

Be very careful to press the tape firmly to the printed circuit material at the point where the tape and pads meet. This is very important, since any air gap left between the tape and copper will permit etchant to enter, resulting in an open circuit and an inoperative printed circuit board.

To lay out the integrated circuits, you may want to obtain IC printed circuit layouts for DIP (dual in-line packages) or a template with which you can accurately locate the IC pins. Since most ICs contain either 14 or 16 pins, it is important that the holes be accurately located. Any inaccuracy in the layout of the IC pins will result in difficulty when attempting to mount the ICs or sockets to the board.

It would be good practice to use sockets for all integrated circuits. It is well worth the extra cost involved since it is extremely difficult to remove a multipin IC from a printed circuit board once it has been soldered in place. Both the printed circuit and the IC could be damaged. The use of sockets has the additional advantage that during test or service of the circuit you can disable one or more pins of the IC by simply bending the pins slightly so that they do not make contact with the socket. This could be a real timesaver in the event that a circuit requires troubleshooting.

The best way to etch the printed circuit is to use a shallow glass tray, such as that used for baking (Fig. 3-4). Best results will be obtained if the etchant is warm, but not hot, during etching. A Pyrex dish can be warmed directly over a stove or hotplate. Be careful not to overheat the etchant.

Place the printed circuit face up in the tray with sufficient etchant to cover the material entirely. You will have to constantly agitate the tray during the etching process, which will take about 20 minutes. If you do not keep the liquid in constant motion the etching time will be increased and it is very possible that some parts of the board will not be completely etched. Be careful to not spill the etchant on your skin or clothing. There is little danger if you happen to do this, but the stains will be very difficult

Fig. 3-4. Etching the direct marked copper-clad board.

to remove. When the board is completely etched remove it from the
solution and wash it thoroughly with plain water. You can save the
etchant for future use. Be sure to thoroughly wash the tray so that no
trace of etchant remains.

Remove the tapes, pads, or resist from the etched board, and polish it
with a piece of fine steel wool. It is very important to remove all traces of
adhesive of oxidation from the copper so that it can be readily soldered.
Failure to do this can result in cold solder connections and an
inoperative circuit.

When drilling the board be sure that the drill bit is sharp. Best results
for glass epoxy board material will be obtained with a carbide tipped
drill, but you can use an ordinary high speed drill bit if that is all you
have. For most components the proper drill size is a No. 57 drill,
0.043-inch (0.109 centimeter) diameter, and for integrated circuits No.
68 drill 0.031-inch (0.079 centimeter) is used. The specified drill sizes
are not critical, but it is best not to use too large a drill since it makes it
more difficult to get a proper solder connection.

For soldering purposes use a light duty iron, about 35 watts, with a
small tip. Be sure the tip is tinned properly and kept clean by wiping it
periodically on a damp cloth or sponge. Use only rosin core solder, and
avoid the use of soldering paste. The use of such paste on electronic

circuits is almost certain to cause a problem in the future if any residue is left on the board. If you have cleaned the copper foil with steel wool as directed you will have no problem in attaining excellent solder joints. Be extremely careful not to overheat the printed circuit, especially around the integrated circuit connections. Excessive heat will cause the copper to lift from the board. If you are inexperienced in soldering printed circuit boards, practice on scrap board material until you can achieve good connections.

If you should inadvertently solder a component into a printed circuit board at the wrong location, it should be removed by drawing the solder away using a "solder sucker" or copper braid solder wick (Fig. 3-5).

Fig. 3-5. Solder sucker bulb and solder wick used for removing solder from printed circuit board.

This technique makes component removal easy and helps prevent damage to the printed circuit board. After the component is removed you can clear the holes, if necessary, using the sucker or wick to allow insertion of the proper component.

Follow the component parts layout shown for the project you are working on. Be sure to mount electrolytic capacitors, diodes, and transistors in the correct direction. When inserting the integrated circuits into the board observe which corner is shown in the component parts layout as pin 1, and then insert the IC so that its pin 1 coincides. ICs are marked with a small dot or numeral 1 molded into the top of the plastic case, or have a U shaped impression molded at the end of the case where pin 1 is located.

After the printed board is completely soldered, give it a thorough visual check to pick up any unsoldered holes, bad solder connections, or solder bridges between adjacent conductors. By following this step you will probably avoid 90% of the problems encountered with home built electronic circuits.

Some of the projects in this handbook use integrated circuits which are not low cost, and in order to protect these chips it would be good practice to apply power to the circuit before plugging the ICs into their sockets. You then will be able to measure the voltage at each and every pin of the IC socket to be sure that no obvious problem exists. The most important voltages to measure are the pins which are connected to the power source, and the pins that are connected to ground. Refer to the schematic diagram of the unit to determine what pins are connected to these points.

Once you have made the above tests and everything seems normal, turn the power source off and you can plug the ICs into their sockets. (Then you can proceed with any test and adjustment of the circuit if necessary.)

If it is necessary to remove an IC from its socket, use an IC puller (Fig. 3-6) that is a very inexpensive tool which will prevent a damaged IC.

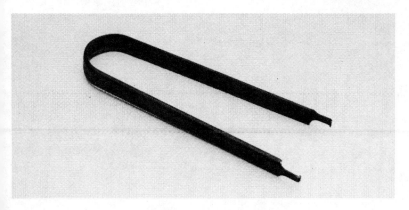

Fig. 3-6. An integrated circuit extractor tool.

Without this tool it is very easy to accidentally bend the IC pins as it is removed from the socket.

If possible measure the current drawn by the circuit, especially in those projects which are battery operated. The text accompanying the project will indicate what range of current is normally being delivered by the battery. Should there be a wiring error or a defective component, the current measurement may alert you that something is wrong.

Chapter 4

Solid-State Device Specifications

This chapter contains information on all solid-state devices used in the construction of the projects in this handbook, with the exception of the integrated circuits. Voltage, current, and gain specifications for each transistor are listed in Table 4-1. Voltage and current specifications for each diode are listed in Table 4-2. These solid-state devices, as used in the projects are not at all critical, and in many cases you will be able to substitute these components if necessary, as long as you adhere to the voltage, current, and power ratings specified. Specifications for diacs, triacs, and LEDs are given in Table 4-3.

In the case of diodes and transistors, there is a wide variety of parts available on the surplus market that can be substituted for those specified in the parts list. If you page through any of the popular electronics magazines you will find many advertisements of sources of supply for electronics parts at reasonable prices. In general, do not substitute germanium diodes for silicon and vice-versa, and adhere to the reverse voltage and forward current specifications for these parts. If the substituted part has a greater voltage rating or forward current specification, it will probably make a satisfactory substitution.

It is not recommended that zener diodes be substituted with others of different voltage ratings. These diodes are designed to produce specific regulated voltages in the circuits, and large variations may cause improper operation. You may substitute a zener diode if it has the same voltage rating as the original diode specified in the parts list.

When substituting transistors, always use a part that has at least the same voltage rating, current rating, and gain as the specified part. You

Table 4-1. Specifications for Transistors

Type Number	Description	Voltage Rating	Maximum Collector Current	DC Current Gain	Gain-Bandwidth Product
2N3904	General purpose npn silicon switching and amplifier transistor	40 Volts	200 mA	100 to 300	300 MHz
2N3906	General purpose pnp silicon switching and amplifier transistor	40 Volts	200 mA	100 to 300	250 MHz
2N4231	Medium power silicon npn transistor	40 Volts	3 A	25 to 100	4 MHz
2N4401	Npn silicon switching and amplifier transistor	40 Volts	600 mA	50 to 150	200 MHz
2N5179	Silicon npn high frequency transistor	12 Volts	50 mA	25 to 250	900 MHz
MPS-A42	Npn silicon high voltage transistor	300 Volts	500 mA	40	50 MHz
MPS-L01	Npn silicon high voltage transistor	120 Volts	150 mA	50 to 300	60 MHz

Table 4-2. Specifications for Diodes

Type Number	Description	Voltage Rating	Maximum Operating Current
1N34A 1N90	General purpose germanium diode	60 Volts	50 mA
1N2069	Silicon rectifier diode	200 Volts	500 mA
1N2070	Silicon rectifier diode	400 Volts	500 mA
1N4148	General purpose silicon switching diode	75 Volts	100 mA
1N4734	Silicon zener diode	5.6 Volts	178 mA
1N5230	Silicon zener diode	4.7 Volts	106 mA
1N5233	Silicon zener diode	6.0 Volts	83 mA

Table 4-3. Specifications for Diacs, Triacs, and Light-Emitting Diodes (LEDs)

Type number	Description
RCA D32024 Motorola 1N5758	Silicon bidirectional diode (diac), 32 volt typical breakdown voltage, 25 microampere breakdown current.
RCA T2500B Motorola 2N6342 GE SC141B	6 ampere, 200 volt silicon triac. Plastic case style TO-220.
Xciton XC209R	Red LED, 2 volts maximum at 20 milliamperes.

must use npn or pnp types as specified. All of the transistors used in these projects are silicon devices, and you should not substitute with germanium. With the exception of the 2N5179 transistor, all transistors used in the projects are rated for audio or switching service. If you must substitute the 2N5179, be sure to use an npn silicon transistor that is rated to operate satisfactorily at the operating frequency of 100 megahertz.

In the case of integrated circuits, you may use any manufacturer's part if it is the exact equivalent of the specified device. Your parts supplier should be able to furnish you with proper substitution information on integrated circuits. The basic digital integrated circuits such as logic gates, counters, timers, and decoders are produced by several manufacturers. These part numbers usually carry the same digits in the part numbers with various letters, depending upon the manufacturer. The more complex integrated circuits may be supplied by only one manufacturer, and it will not be possible to substitute these items. For complete information on any of the integrated circuits used in the projects of this handbook, refer to the manufacturer's specification sheets for these chips.

Section 2

Electronic Telephone Projects

Chapter 5

Soft Tone

The Soft Tone bell (Fig. 5-1) is the simplest project in this handbook, and might be a good first project for those who have never built any telephone accessories before. It will be useful wherever the normal telephone bell is either too loud or annoying, and it also is a valuable addition to another project in this book, *Intercom Phone*. It also can be used as a remote addition to the existing telephone bell. Such use permits the telephone ringing signal to be provided wherever you care to run a pair of wires. Soft Tone does exactly what its name implies—it

Fig. 5-1. The Soft Tone bell

produces a pleasant, soft, musical tone for each ring of the telephone. It would be an ideal accessory for a telephone located in a bedroom, for example. Since it is designed around the solid state Mallory Mini Sonalert module, the builder has the option of selecting from a variety of frequencies and sound levels. The Mini Sonalert specified in the parts list will produce a very soft and pleasing tone of 3500 hertz, but other modules may be used for greater sound intensity and different operating frequencies. Fig. 5-2 is a chart listing three Sonalert units which are

TYPE NUMBER	FREQUENCY	SOUND LEVEL	LOUDNESS
SC18	3500 HERTZ	70 dB	SOFT
SC628	2900 HERTZ	80 dB	MEDIUM
SC616N	2900 HERTZ	95 dB	LOUD

Fig. 5-2. Recommended Mallory Sonalert modules.

suitable for telephone use. The use of a solid-state signaling device for this application provides the utmost in reliability, since there are no moving parts to wear out.

ABOUT THE CIRCUIT

As shown in the schematic diagram (Fig. 5-3), Soft Tone consists of just three parts. The solid-state electronic signal module is operated by the 90-volt 20-Hz ringing signal which appears across the telephone line when the phone rings. The resistor and diode have been included in the circuit to limit the current fed to the module, and protect it against the reverse voltage which appears across the line during each half-cycle of the ringing signal. The diode also presents an open circuit condition to the telephone line when it is not in use. This feature makes Soft Tone undetectable when connected across the line, and will not affect telephone performance in any way.

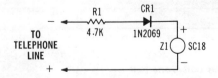

Fig. 5-3. Schematic diagram of the Soft Tone.

CONSTRUCTION

The parts list is given in Table 5-1. Refer to Fig. 5-4 which is a full size layout of the printed circuit board used in the construction of soft tone. Fig. 5-5 illustrates the component layout as seen from the top of the board. The printed circuit layout shown is for use with the 3500-Hz Sonalert specified in the parts list. Drill 3/16 inch (4.76 millimeter) diameter holes for the tabs of the Sonalert, and solder it into place by

Table 5-1. Soft Tone Project Parts List

Item	Description
CR1	Silicon diode, 100-mA, 200-volt PIV, 1N2069 or equivalent
R1	4.7K, ½-watt, 10% resistor
Z1	Sound module, Mallory SC18 or equivalent

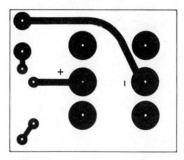

Fig. 5-4. Printed circuit layout (foil side).

tacking each tab to the copper foil. If any of the other Sonalert modules are to be used you will have to modify the layout so that the + and − terminals of the module fit into the drilled holes in the printed circuit board. Be sure to observe correct polarity of the Sonalert module and diode as shown. The value of R1 can be increased to provide less sound intensity if desired, but do not use a value lower than that specified.

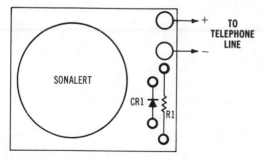

Fig. 5-5. Component layout (component side).

INSTALLATION

Soft Tone is small enough to be concealed inside the ordinary K500 telephone. If your telephone is one of the newer styles which are slim and compact you will have to locate the module outside the telephone case.

If you wish to disconnect the existing telephone bell, refer to Fig. 5-6 which is a simplified diagram illustrating the interior connections of

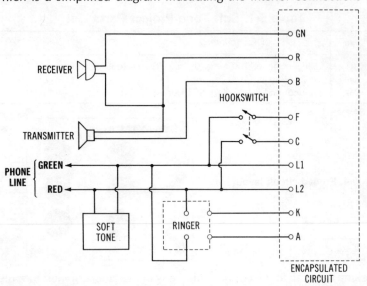

Fig. 5-6. Simplified internal telephone circuit showing connections for Soft Tone.

most telephones. Note that for private telephone lines the telephone bell is connected across L1 and L2, the green and red telephone line wires. If you happen to have a two party line the bell will be connected between either L1 or L2, and ground. The coil of the ringer is also connected to terminals A and K of the encapsulated assembly within the telephone. Disconnecting any one of the wires of the ringer from terminals L1, L2, A or K will render it inoperative. If desired, you can add a single pole switch to the bell circuit so that it can be easily restored to normal operation if ever necessary.

When connecting Soft Tone across the telephone line (L1 and L2), you will have to observe correct polarity. This is very easily done, for if you do connect the wires incorrectly, Soft Tone will emit a continuous tone, responding to the 48 volts dc which is ever present across the telephone line. If this should occur, simply reverse the two connections and the installation will be completed.

40

Chapter 6

Telephone Sentry

In many homes today there is more than just one telephone connected to the line. It is quite common to have one or more extension phones for convenience. But, along with that convenience is a distinct disadvantage. It is possible that another person could surreptitiously listen in on a telephone conversation without either party knowing it if he or she is very careful when lifting the receiver so that no sound is produced. This problem has always existed in two or more party lines, many of which are still in use today. This is where Telephone Sentry (Fig. 6-1) comes in. With this circuit attached to the telephone it will be

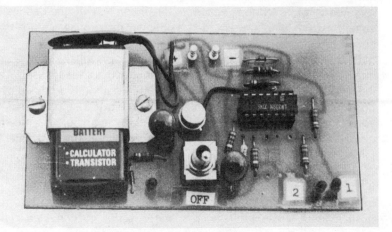

Fig. 6-1. The Telephone Sentry.

possible to detect the presence of one or two additional telephones connected to the line. It uses a common 9-volt transistor battery and is easy to connect to the telephone line. Current draw from the battery is in the 10 to 20 milliampere range which permits long battery life. The circuit presents a high impedance to the telephone line and will not affect telephone performance.

The Telephone Sentry uses two LED lamps to indicate the presence of additional telephone extensions in use. Under normal conditions both LEDs will be extinguished. Should one extension be picked up, one LED will light. The second LED will light if another extension is picked up. A third LED is used to provide an indication of battery condition. As long as the battery has sufficient terminal voltage to properly operate the unit, LED 3 will be illuminated. As the battery nears the end of its useful life the intensity of LED 3 will decrease dramatically, indicating that the battery needs replacement.

ABOUT THE CIRCUIT

The key to the operation of the Telephone Sentry is the dc voltage which appears across the telephone line. As explained in Section 1 of this handbook the telephone line acts like a battery of about 48 volts dc behind a source impedance of about 1000 ohms. When one telephone is connected across the line the dc voltage drops to about 4.6 volts. When a second telephone is added across the first, the voltage drops to about 3.2 volts, and when a third telephone is connected the voltage drops to about 2.3 volts. Although these voltage levels will vary somewhat from telephone system to telephone system, the change in voltage as each telephone is added is sufficient for a voltage comparator circuit to detect.

The heart of the Telephone Sentry is a four section voltage comparator integrated circuit, LM339. Only two sections of this chip are used. The reference voltage for each voltage comparator section is provided by a nonadjustable 5-volt regulator integrated circuit, IC2, and two sets of voltage divider networks. By using a voltage regulator to hold the reference voltage constant, the problem of unreliable operation due to falling battery voltage is avoided.

The voltage detection level has been set at 3.9 volts for section A of IC1, and 2.75 volts for section B. These voltages are fed to the negative inputs of each voltage comparator section, pins 6 and 8, of IC1. The positive input of each voltage comparator section is connected to the telephone line through a 47K resistor which provides a high impedance to the telephone line. The outputs of the comparators, pins 1 and 14 of IC1, will assume a voltage of either zero or five volts, depending upon the voltage across the telephone line. If only one telephone is connected across the line, the voltage fed to the positive inputs of the two comparator sections will be 4.6 volts, which is greater than 2.75 and 3.9 volts. This will cause the outputs of the comparators to be 5 volts. Thus neither LED will be illuminated. Should one extension be picked up, the voltage across the line will be more than 2.75 volts, but less than 3.9 volts. This will cause the output of IC1A to become zero volts, and illuminate LED1. When another extension telephone is added across the line, the voltage across the line will be less than 2.75 volts. This causes the output of IC1B to go to zero volts and LED2 becomes illuminated.

CR1 has been placed in the circuit to prevent the possibility of the 48-volt telephone line voltage from damaging the circuit should it be left on when no telephones are in use. A double pole power switch is used to disconnect both the battery and the telephone line when the Telephone Sentry is not in use.

CONSTRUCTION

The parts list is given in Table 6-1. The Telephone Sentry is constructed on a 2½ × 4½ inch (6 × 11 cm) printed circuit board as shown in the photograph of Fig. 6-1, and the full size printed circuit layout (Fig. 6-2). Fig. 6-3 shows the component layout. Although the model has been constructed with the LED indicators and power switch mounted to the printed circuit board, the builder may want to house the assembly in a small cabinet with the LEDs and switch mounted to the cabinet. The schematic diagram of the circuit is shown in Fig. 6-4.

It is recommended that a 14-pin IC socket be used for IC1. This will allow the integrated circuit to be removed from the printed circuit board for service if it ever should be necessary. Once a multipin IC has been soldered into a printed circuit, it is extremely difficult to remove without destroying the printed circuit or IC. Be sure to mount IC1 in the correct direction as shown in the components parts layout. Pin one of IC1 is indicated by a small dot on the parts layout and printed circuit layout. The IC itself has pin 1 identified by a small dot or number 1 molded in the top of the plastic case, or a groove at one end. When mounting the LED indicators, diode CR1, and the electrolytic capacitors follow the correct polarity shown in the parts layout.

Table 6-1. Telephone Sentry Parts List

Item	Description
C1	22-μF, 15-volt electrolytic capacitor
C2	22-μF, 15-volt electrolytic capacitor
R1	12K, ¼-watt, 10% composition resistor
R2	47K, ¼-watt, 10% composition resistor
R3	470-ohm ¼-watt, 10% composition resistor
R4	39K, ¼-watt, 10% composition resistor
R5	47K, ¼-watt, 10% composition resistor
R6	470-ohm, ¼-watt, 10% composition resistor
R7	47K, ¼-watt, 10% composition resistor
R8	270-ohm, ¼-watt, 10% composition resistor
CR1	Silicon diode, 75-volt PIV, 1N4148 or equivalent
CR2	Zener diode, 4.7-volt, 1N5230 or equivalent
IC1	Integrated circuit, LM339 National Semiconductor or equivalent
IC2	Integrated circuit, LM309H National Semiconductor or equivalent
SW1	Switch, dpdt, C & K 7201 or equivalent
LED1	Light-emitting diode, Radio Shack 276-090 or equivalent
LED2	Light-emitting diode, Radio Shack 276-090 or equivalent
LED3	Light-emitting diode, Radio Shack 276-090 or equivalent
B1	9-volt transistor battery, Eveready 1222 or equivalent
Battery Clip	Herman H. Smith No. 1234 or equivalent

One wire jumper is used in the circuit as shown in the component parts layout. Use a piece of insulated wire for the jumper to avoid the possibility of a short circuit to any of the components.

The battery is mounted to the printed circuit board by a clamp formed out of a strip of copper about 1-inch (2.54 cm) wide, and two No. 4 machine screws. The ends of the battery cable are soldered directly into the printed circuit board. Again, be sure to follow the correct polarity.

CIRCUIT TEST

The Telephone Sentry can be tested very easily before installation by turning on the power switch. LED3 should be illuminated indicating that the battery voltage is sufficient to properly operate the circuit. Short the telephone line connections, L1 and L2, together. Both LED1 and LED2 should light. If desired, the voltage detection levels at pins 6 and 8 of IC1 can be measured with a vtvm. This information will help if any adjustments of the circuit values are necessary due to variations in the particular telephone line the unit is to be installed on.

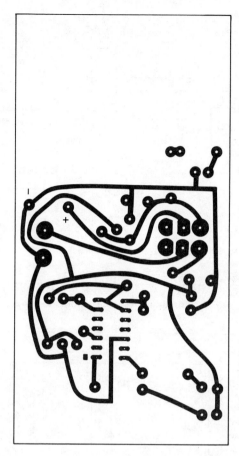

Fig. 6-2. Printed circuit layout shown full size (foil side).

INSTALLATION

Before connecting the Telephone Sentry to the telephone line you must determine the polarity of the dc voltage on the line. This can be done with any dc voltmeter with a voltage range of at least 50 volts. Once the polarity of the telephone line is known, the Telephone Sentry can be connected across the line, L1 to positive, and L2 to negative.

To check circuit operation, measure the dc voltage across the telephone line when two and three telephone extensions are in operation. This information can be used, if necessary, to tailor the resistance values of R1 and R4 so that the voltage detection levels fed to pins 6 and 8 of IC1 are about midway between the actual operating

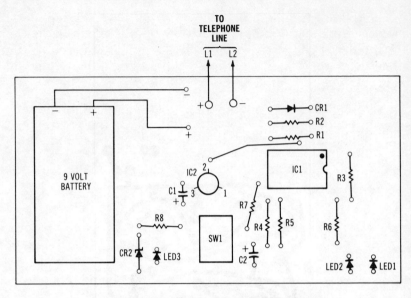

Fig. 6-3. Component layout (component side).

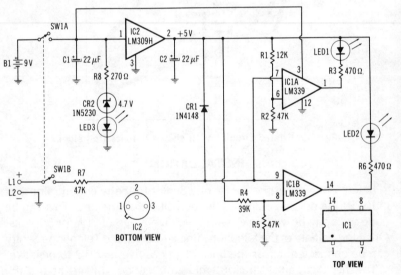

Fig. 6-4. Schematic diagram of the Telephone Sentry.

voltage levels of your telephone line. The voltage detection levels for
the resistance values specified in the parts list are 2.75 and 3.9 volts.
These levels were used for actual voltage operating levels of 4.6, 3.3,

and 2.3 volts measured on a particular telephone line. If the operating voltage levels of your telephone line differ by more than 0.3 volt, you may want to change the values of R1 and R4 for a better match to your telephone system.

Chapter 7

Telephone Bug

There are many ways to "bug" a telephone but this tiny transmitter (Fig. 7-1) is probably one of the simplest and easiest methods with which to eavesdrop on a telephone conversation. It consists of a one-transistor Colpitts oscillator which derives its power from the telephone line. Since it places a resistance of less than 100 ohms in series with the telephone line, it has no effect on telephone performance nor can it be detected by the telephone company or anyone using the telephone. Since this is a very low power device using no antenna for radiation, it can be considered to operate under the requirements of part

Fig. 7-1. The telephone Bug.

15 of the FCC Rules and Regulations. Part 15 allows unlicensed devices to be used providing that the electromagnetic field produced at a specified distance from the transmitter be not greater than 15 microvolts per meter. Note also that Part 15 of the regulations also specifies that no person shall use such a device for eavesdropping unless authorized by all parties of the conversation. Installation is accomplished by connecting the Bug in series with the telephone line at any point. An ordinary fm radio is used as the receiver, and both sides of the telephone conversation can be heard.

The entire circuit is mounted on a printed circuit board measuring 2 inches × 2 inches (5 cm × 5 cm) without any attempt to make the unit as small as possible. Fig. 7-2 is a full size layout of the printed circuit as

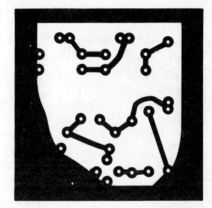

Fig. 7-2. Printed circuit layout shown full size (foil side).

seen from the foil side of the printed circuit board. The small size of this transmitter makes it easy to conceal, if necessary. Fig. 7-3 is a view of the component side of the printed circuit board.

ABOUT THE CIRCUIT

The schematic of the Bug is shown in Fig. 7-4. One side of the telephone line is broken at any point, and the two input terminals of the Bug are connected to the exposed ends of the broken telephone line. A 100-ohm resistor, R1, completes the telephone circuit so that normal telephone performance is not affected. As mentioned in Chapter 1, the impedance of the telephone circuit is much greater than 100 ohms, and the addition of another 100 ohms into the circuit will have no effect.

Diodes CR1, CR2, CR3, and CR4 form a bridge rectifier that produces a varying dc voltage according to the audio signals on the telephone

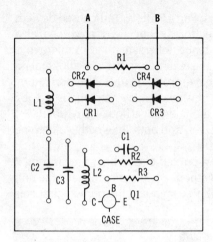

Fig. 7-3. Component layout (component side).

line. The full-wave bridge circuit ensures that a dc voltage will be produced regardless of the polarity of the varying dc on the telephone circuit. This dc voltage is used as the supply voltage for the oscillator transistor, Q1.

Transistor Q1 is connected as an ordinary common-emitter Colpitts oscillator. The tuned circuit for this oscillator is composed of C2, C3, and L2. The collector of Q1 is connected to one end of the tuned circuit, and the base is connected to the other end. The center tap of the tuned

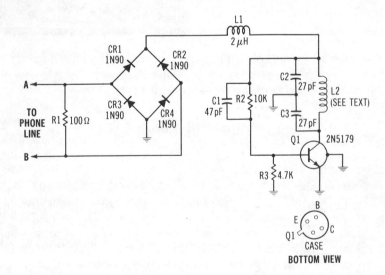

Fig. 7-4. Schematic diagram of the telephone Bug.

circuit, provided by the series connection of C2 and C3, is connected to the emitter of Q1. This provides the necessary feedback from collector to base and causes Q1 to oscillate at the resonant frequency of L2, C2, and C3. The circuit oscillates at approximately 95 MHz which is within the range of the standard fm band.

Since the dc voltage fed to the collector and base of Q1 is derived from the varying audio current in the telephone line, the oscillation frequency of Q1 is not stable and varies in accordance with the audio current of the telephone line. Thus Q1 is a modulated oscillator whose frequency contains audio information. Q1 actually produces both amplitude and frequency modulation, but the fm receiver used to pick up the signal will respond only to the frequency modulated component of the rf.

CONSTRUCTION

The parts list is given in Table 7-1. The printed circuit board layout and component parts layout are shown in Figs. 7-2 and 7-3. When mounting the diodes be sure to follow the exact polarity as shown. L2 consists of five turns of number 20 solid wire wound on a ¼-inch form. Once L2 is wound, it will not require any form. Simply solder it into the printed circuit board, which will hold it nicely in place. The parts list specifies a 2-microhenry radio frequency choke for L1. This part may be constructed by winding approximately 40 turns of number 36 enamel wire on a one-megohm, ½-watt resistor. The value of L1 is not critical. When mounting Q1, solder it into the printed circuit board leaving about ¼ inch or less lead length. This circuit operates at approximately 95 MHz, and it is always good practice to use short lead lengths at this frequency. Be sure to position Q1 so that the leads are inserted into the proper holes.

Table 7-1. Telephone Bug Parts List

Item	Description
C1	47-pF ceramic disc capacitor
C2, C3	27-pF mica capacitor
CR1, CR2, CR3, CR4	Germanium diode, 1N90 or equivalent
R1	100-ohm, ¼-watt, 10% composition resistor
R2	10K, ¼-watt, 10% composition resistor
R3	4.7K, ¼-watt, 10% composition resistor
L1	2-μH radio frequency choke (see text)
L2	5 turns No. 20 gauge wire (see text)
Q1	Npn rf transistor, 2N5179 or equivalent

TEST AND ADJUSTMENT

It would be good practice to check the Bug on the workbench before installing it on the telephone line. To do this you can use a 6.3-volt filament transformer as the power source and an fm radio as the receiver. Connect the Bug to the 6.3-volt winding of the transformer, as shown in Fig. 7-5 and tune the fm radio over the entire fm band until you hear the 120-Hz modulation produced by the Bug. It may be necessary to increase or decrease the inductance of L2 slightly to bring the frequency of oscillation within the range of the fm receiver. This may be accomplished by squeezing the turns of L2 closer together, or pushing them farther apart. It it best to adjust the frequency of oscillation in this manner so that it does not coincide with a strong fm station. Once this adjustment is made the Bug can be installed on the telephone line.

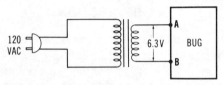

Fig. 7-5. Test setup for telephone Bug.

INSTALLATION

To place the Bug in operation it will be necessary to break one side of the telephone line. The easiest place to do this is at the terminal block located at the end of the telephone wire coming into the house from the telephone pole. This block is usually placed in the basement of the home. Simply remove one of the two wires from the terminal block and connect the Bug between the open wire and the terminal as shown in Fig 7-6. It is not necessary to observe polarity, since the Bug will operate with current flowing in either direction. Once the connections are made, the Bug will automatically transmit both sides of the telephone conversation.

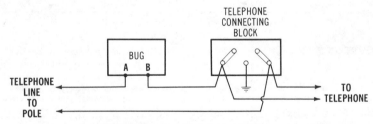

Fig. 7-6. Typical installation wiring.

Chapter 8

Automatic Record/Call

If you own a tape recorder you have more than half of an automatic telephone recording system. All you need is a little electronics as described in this project and you will be able to have an automatic telephone recording system which will put on tape all incoming and outgoing calls from your telephone. The only attention the system will require is a periodic check to see if the tape has been used up. If your tape recorder is the type which has a remote microphone you will be able to connect Record/Call (Fig. 8-1) to your recorder without making any modifications whatsoever. Tape recorders which do not have this feature will require a very simple modification to bring out a pair of wires for the motor and audio input. In either case the tape recorder will be easy to disconnect from the recording system and will be free to be used for other purposes.

Due to the wide variety of tape recorders available on the market today it will be necessary that you determine the polarity of the battery or power supply in your tape recorder, and then follow the circuit diagram which pertains to your unit. Two circuits are provided, one for negative ground systems, and one for positive ground systems. If your tape recorder was supplied with a schematic diagram you can determine the power supply polarity by noting which side of the battery is connected to the chassis. An alternate method would be to use a dc voltmeter. Connect the negative lead of the voltmeter to any metal portion of the tape recorder and connect the positive lead first to one side of the battery and then the other. If you get a positive voltage reading at the positive end of the battery you have a negative grounded system. Conversely, if you read a negative voltage at the negative end of the battery and zero volts at the positive end, you have a positive grounded system. Regardless of which system you have, the printed circuit layout for Record/Call will be the same, but the polarity of the transistors, diodes, and electrolytic capacitor will be different.

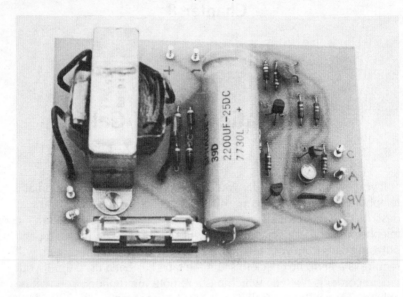

Fig. 8-1. Automatic Record/Call.

ABOUT THE CIRCUIT

The electronic circuit for Record/Call consists of three sections—power supply, solid-state motor control, and audio interface. An ac operated power supply has been included in this unit for two reasons. It may be desired to leave the power on for long periods of time, in which case battery operation would not be practical. Also, the built-in power supply could also be used to power the tape recorder, provided it will operate with a 9-volt supply. Thus you would be able to operate the recording system without the necessity of using batteries at all. If recording is required for only short periods of time, you have the option of deleting the ac power supply entirely and using a common 9-volt transistor battery to power the circuit. The battery power supply option is shown in the schematic diagrams with dotted lines. Bear in mind if you use a 9-volt transistor battery it can power only the printed circuit board, not the tape recorder motor.

The ac power supply is a common 6.3-volt filament transformer feeding a full-wave bridge rectifier. The output of the rectifier is smoothed to approximately 9 volts dc by C1. The value of C1 is large enough to provide sufficient filtering to operate the tape recorder and motor, if so desired. The output of the power supply feeds the solid-state motor switch and the audio section of the unit.

The signal which operates Q1 and Q2, the motor switch, comes from the telephone line. When the telephone is idle, the voltage across the line is about 48 volts dc. This voltage is coupled to the base of Q1 through a voltage divider composed of R1 and R2, and causes Q1 to be cut off. The collector current of Q1 (and base current of Q2) is zero. Thus Q2 is also cut off. Since the collector of Q2 is connected to the return lead of the tape recorder motor, the motor sits idle. When the telephone line is in use, the voltage across the line drops to about 5 volts. The combination of a 5-volt voltage source through R1 and the forward bias through R2 is sufficient to saturate Q1, resulting in forward bias to Q2 through R3. Q2 turns on like a switch, completing the circuit of the tape recorder motor which operates the recorder in the normal way. A diode has been placed between the base and emitter of Q1 to prevent damage to the circuit from the 90-volt ringing signal impressed across the telephone line.

The remaining circuit on the printed circuit board is the interface section between the telephone line and the audio input of the tape recorder. The purpose of this circuit is to detect the audio information on the telephone line and feed it to the relatively low impedance input of the tape recorder without placing a substantial load on the telephone line. Since the carbon microphone used in telephone service has a much greater voltage output than the typical tape recorder microphone, no amplication of the audio signal is necessary. Q3 is connected as an emitter-follower to provide the high to low impedance transformation. R6, the emitter resistor of Q3, is a potentiometer which allows a convenient method of setting the amplitude of the audio signal fed to the tape recorder for best audio fidelity. Q3 is isolated from the dc voltage on the telephone line by C2, and responds only to the audio voltage fed to its base.

CONSTRUCTION

The parts list is given in Table 8-1. Record/Call is constructed on a 3½-inch × 5-inch (9 cm × 13 cm) printed-circuit board. The printed circuit is shown full size in Fig. 8-2. Before constructing the board check

Table 8-1. Automatic Record/Call Parts List

Item	Description
C1	2200-μF, 25-volt electrolytic capacitor
C2	.01-μF ceramic disc capacitor
C3	0.1-μF ceramic monolithic capacitor
CR1, CR2, CR3, CR4	Silicon diode, 500-mA, 100-volt PIV, 1N2069 or equivalent
CR5	Silicon diode, 75-mA, 100-volt PIV, 1N4148 or equivalent
R1	100K, ¼-watt, 10% composition resistor
R2	47K, ¼-watt, 10% composition resistor
R3	470-ohm, ¼-watt, 10% composition resistor
R4	220K, ¼-watt, 10% composition resistor
R5	220K, ¼-watt, 10% composition resistor
R6	5K potentiometer
R7	47K, ¼-watt, 10% composition resistor
T1	6.3-volt, 1-amp filament transformer
Q1	(Negative ground) pnp silicon transistor, 2N3906 or equivalent
Q1	(Positive ground) npn silicon transistor, 2N3904 or equivalent
Q2	(Negative ground) npn silicon transistor, 2N3904 or equivalent
Q2	(Positive ground) pnp silicon transistor, 2N3906 or equivalent
Q3	(Negative ground) pnp silicon transistor, 2N3906 or equivalent
Q3	(Positive ground) npn silicon transistor, 2N3904 or equivalent
F1	¼-amp, 250-volt fuse, Type 3AB or equivalent
Fuseholder	Buss 4405 or equivalent

the full size layouts as shown in Figs. 8-3A or 8-3B to be sure that the transformer that you are going to use will fit in the space provided. If your transformer is larger than that shown it is a simple matter to increase the size of the board as necessary. Note that the ac power line circuit is protected by a ¼-ampere fuse mounted in a fuseclip assembly on top of the printed circuit board. Use a three prong ac plug for powering the transformer, and wire the leads so that the hot lead of the ac power line is connected to the fuse.

The value of C1 as specified in the parts list is great enough to satisfactorily operate the tape recorder motor from the power supply. If you are going to operate the tape recorder from its own power source, the value of C1 can be reduced to about 100 μf.

Before mounting the diodes, transistors, and electrolytic capacitor on the printed circuit board, check the power supply or battery polarity of the tape recorder you are going to use as directed in the beginning of the text. Follow only that component parts mounting diagram which pertains to your unit. If you inadvertently use the wrong diagram the circuit will not work.

Connections can be made to most tape recorders by the use of miniature plugs which are readily available from electronics supply houses, such as Radio Shack, Lafayette, etc. If your tape recorder does not have provision to control the motor from a remote microphone you can easily modify it by opening the motor lead which is connected to the chassis and adding a closed circuit jack to the unit. A typical wiring connection for this plug through which the motor control wire from the tape recorder is connected to the collector of Q2 is shown in Fig. 8-4. A

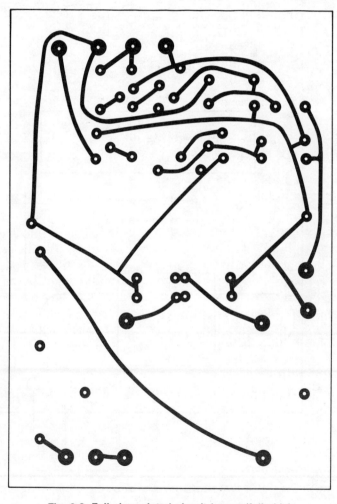

Fig. 8-2. Full size printed circuit layout (foil side).

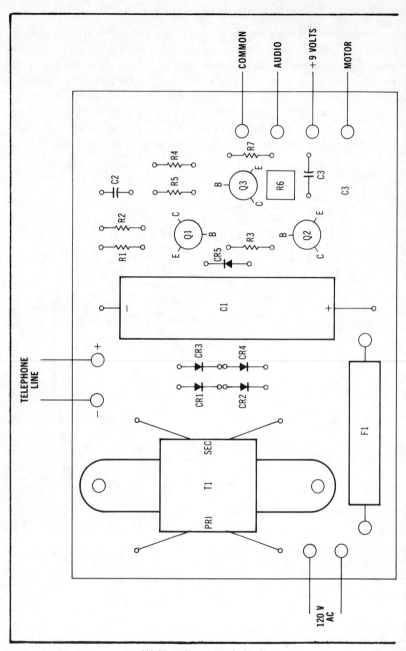

(A) Negative ground circuit.

Fig. 8-3. Component layout

58

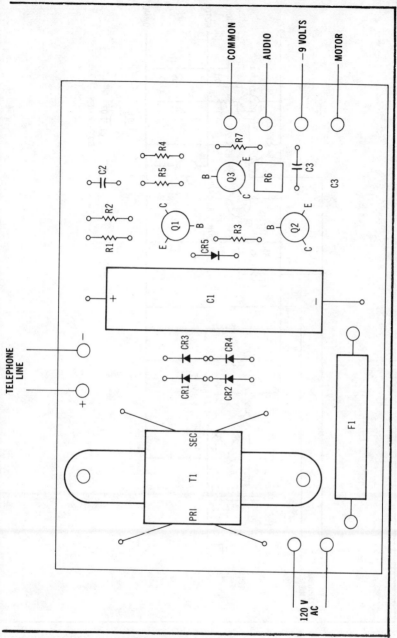

(B) Positive ground circuit.

(component side).

59

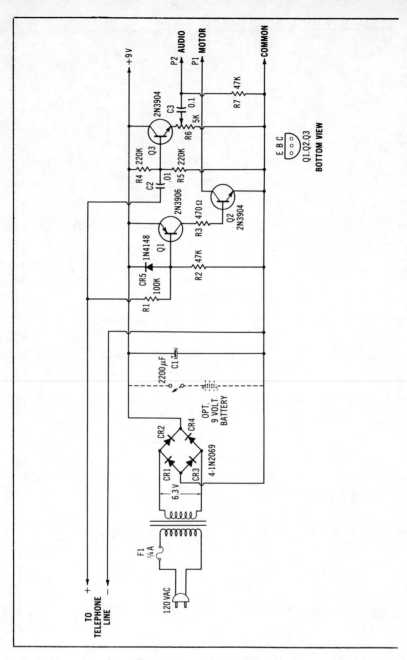

(A) Negative ground recorders.

Fig. 8-4. Schematic diagrams

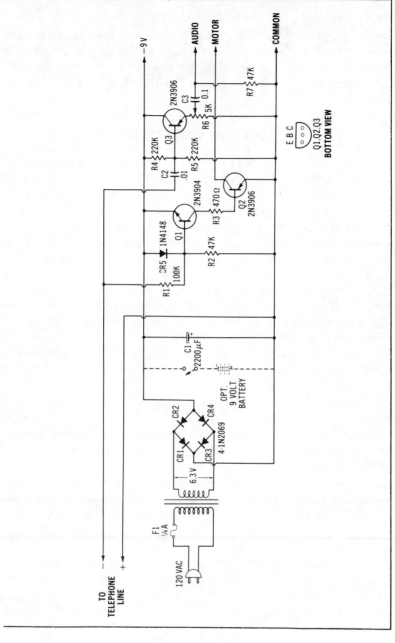

(B) Positive ground recorders.

for the Record/Call.

second plug, usually a different size than the first one, can be used to connect the audio output of Q3 to the microphone input jack of the tape recorder. By using these two plugs for the connections, you can easily disconnect the tape recorder from the recording system when so desired.

TEST AND ADJUSTMENT

Before making any connections to the tape recorder and telephone line the printed circuit should be checked for proper operation. To do this, apply 120 volts ac power to the ac input terminals and measure the voltage across C1, which should be approximately 9 volts dc. If you have an adjustable dc power supply which can deliver about 48 volts, you can check the operation of the solid-state switch. Connect the motor control plug to the tape recorder and apply 48 volts dc to the telephone line input terminals of the printed circuit board, observing correct polarity as shown in the schematic which applies to your tape recorder. Turn the tape recorder on and set its controls to record mode. The motor should not run at this time. Reduce the voltage fed to the telephone line terminals to approximately 5 volts. The motor should now run. This completes the bench test of Record/Call. Audio level, the final adjustment, will be made under actual operating conditions.

INSTALLATION

Before connecting the printed circuit board to the telephone line you must first determine the dc polarity of the telephone line. To do this you can use an ordinary vtvm or vom. The polarity of the telephone line can be checked by setting the meter to dc operation, 50-volt range, and connecting it across the telephone line. When you have identified the polarity of the telephone line the printed circuit board telephone input terminals can be connected as shown in the schematic diagram. Be sure to observe correct polarity as specified by the schematic diagram which pertains to your recorder. The positive side of the telephone line is to be connected to the positive input terminal of the printed circuit, and negative to negative.

Connect the two plugs wired to the output terminals of the printed circuit board to the tape recorder, and turn the recorder to record mode. Set R6, the audio gain control, to about midposition for an approximate setting. When the telephone receiver is lifted off the hook the tape recorder motor should run, recording any signals on the line. Dial a telephone number of some recorded message, which will be recorded on your tape. When the call is completed and the receiver is placed back on the hook, the motor will stop. Rewind the tape, play the recording, and note if the audio sounds too loud and distorted or too soft. You can adjust R6 accordingly, and record another telephone call. After one or two adjustments of R6 you should be able to set the audio gain to a point which gives reasonable fidelity to the recording.

Chapter 9

Intercom Phone

Intercom Phone (Fig. 9-1) is actually a telephone set that is used to initiate and receive telephone calls. You talk into it and hear the other party in the same manner as if you were using an intercom. The advantage of this unit is that you will be able to hold a telephone conversation without the requirement of holding a handset to your ear. Also, other people in the room will be able to hear the conversation of the person at the other end and join in if they desire to do so. If you add two other projects described in this handbook, Soft Tone and Computer Memory, you will have a complete phone set which will ring and from which you can make outgoing calls. Intercom Phone is a complete self-contained unit operated from its own 9-volt transistor battery that can be operated wherever you care to run a pair of telephone wires.

HOW IT WORKS

In order to operate a unit such as this with a microphone and loudspeaker mounted next to each other in the same cabinet, a special technique must be used to avoid feedback. An ordinary telephone is able to operate in both directions, send and receive, at the same time. As this is done with a special transformer, no feedback problem occurs because the receiver and microphone are acoustically isolated. Intercom Phone uses no transformer and has been designed to switch rapidly back and forth between send receive modes according to the voice signals within the unit. This means that one party must stop talking in order to allow the circuit to switch over so that the other party can be

Fig. 9-1. The Intercom Phone.

heard. The operation of the circuit is simple and straightforward.

Instead of using a hybrid transformer such as is used in the telephone, Intercom Phone makes use of a modified Wheatstone bridge to provide isolation between the send signal from the microphone amplifier and the receive signal from the telephone line. Fig. 9-2 is a block diagram of the unit and will help in understanding how the circuit operates. The bridge circuit as shown in Fig. 9-2 is composed of C5, R7, R8, R9, and the telephone line itself. Since the telephone line is not purely resistive at voice frequencies, the balancing network of C5 and R7 placed in the conjugate arm of the bridge provides a reasonable bridge balance for the telephone line. Under this condition, any signals fed to the junction of C5 and R9 and IC1B will be impressed across the telephone line and will be attenuated at the input terminals of differential amplifier IC3A. When signals from the telephone line drive the bridge, the circuit is not in balance, and the signal levels at each input of differential amplifier IC3A will not be equal. It is the difference in the amplitude of the signals at the output of IC1B and IC3A which determines the send or receive mode of Intercom Phone.

IC1A is a voltage amplifier which converts the relatively low output of the microphone to a level which is sufficient to drive the telephone line, about 1 volt peak-to-peak. The gain of IC1A is about 100, and is determined by the ratio of R4 and R27 (Fig. 9-3). C2 is used as a

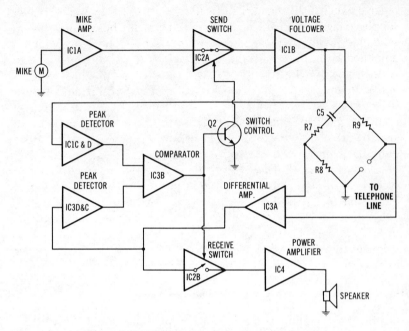

Fig. 9-2. Block diagram of the Intercom Phone.

frequency rolloff capacitor to attenuate frequencies above the normal operating range of the telephone. The output of IC1A is fed to send switch IC2A.

The circuit is normally held in send mode by means of a small positive bias fed to IC1C through R13. This keeps send switch IC2A closed and permits the output of IC1A to be fed to the bridge circuit through voltage follower IC1B. Thus, any sounds reaching the microphone will be sent to the telephone line. The output of IC1B is also fed to the send peak detector circuit composed of IC1C and IC1D.

In a similar manner the output of IC3A is fed to the receive peak detector circuit composed of IC3D and IC3C. Both peak detector circuits are identical except for the time constants involved, and both develop a dc voltage at the output which is essentially equal to the peak of the signal fed to the input. When Intercom Phone is in the send mode, the output of the send peak detector will always be greater than the output of the receive peak detector due to the attenuation chracteristic of the bridge circuit.

Voltage comparator IC3B compares the outputs of the peak detector circuits and controls the send and receive switches. When the unit is in send mode the output of the comparator goes to zero volts. This opens receive switch IC2B and cuts off switch control Q2. R25 applies power supply voltage to pin 13 of IC2A, keeping it closed. Thus the unit stays in send mode as long as there is no signal from the telephone line which is amplified by IC3A.

Any signal coming from the telephone line will be amplified by IC3A and fed to the receive peak detector. As soon as the output of the peak detector exceeds 0.1 volt, the dc bias fed to IC1C, the output of comparator IC3B switches to about 9 volts. This closes the receive switch IC2B and saturates Q2. Send switch IC2A opens, cutting off the output of the microphone amplifier. In this manner the received signal always has priority over the operation of the unit, and will hold it in receive mode as long as a receive signal is present at the input of differential amplifier IC3A. The receive peak detector time constant is determined by R17 and C10 and is necessary so that the circuit remains in receive mode between words and syllables.

Power amplifier IC4 can deliver about 300 milliwatts of audio power to a loudspeaker. This will produce sufficient volume to fill a room with sound. The input to IC4 is controlled by potentiometer R26 which is used as a volume control.

Q1 together with R10, C18, and R11 is used as dc load on the telephone line when Intercom Phone is in use. The dc current drawn by this circuit is used by the telephone company equipment to indicate that the circuit is in use. Q1 presents a very high ac impedance across the telephone line so that the bridge balance is not affected.

Power for the unit is supplied by a common 9-volt transistor battery. The control switch for the unit controls battery power and also disconnects one side of the telephone line when Intercom Phone is not in use. The current drain on the battery is about 20 milliamperes at low volume output, and increases as greater volume is used. If Intercom Phone is operated at normal volume levels the battery should provide reasonably long life. It is not recommended that an ac power supply be used to power this unit, since any 60-Hz hum appearing in the circuit could upset the automatic switching circuitry.

CONSTRUCTION

The parts list is given in Table 9-1. The electronics circuit for Intercom Phone is constructed on a printed circuit board measuring 4⅜ inches × 5 inches (11 cm × 13 cm). The foil layout for this board is shown full size in Fig. 9-4 and the component layout is shown in Fig. 9-5. It is strongly recommended that sockets be used for the integrated circuits.

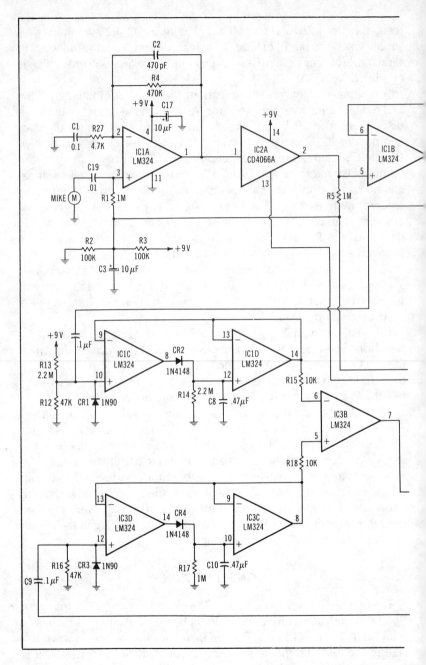

Fig. 9-3. Schematic diagram

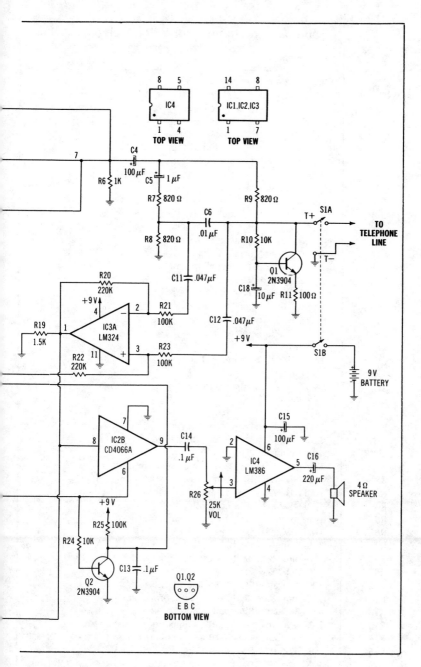

of Intercom Phone.

Table 9-1. Intercom Phone Parts List

Item	Description
C1	0.1-μF, 25-volt ceramic capacitor
C2	470-pF, 25-volt ceramic disc capacitor
C3	10-μF, 10-volt tantalum capacitor
C4	10-μF, 10-volt tantalum capacitor
C5	1-μF, 10-volt tantalum capacitor
C6	0.01-μF, 25-volt ceramic disc capacitor
C7	0.1-μF, 25-volt ceramic capacitor
C8	0.47-μF, 25-volt ceramic capacitor
C9	0.1-μF, 25-volt ceramic capacitor
C10	0.47-μF, 25-volt ceramic capacitor
C11	0.047-μF, 50-volt tubular capacitor
C12	0.047-μF, 50-volt tubular capacitor
C13	0.1-μF, 25-volt ceramic capacitor
C14	0.1-μF, 25-volt ceramic capacitor
C15	100-μF, 10-volt tantalum capacitor
C16	220-μF, 10-volt tantalum capacitor
C17	10-μF, 10-volt tantalum capacitor
C18	10-μF, 10-volt tantalum capacitor
C19	0.01-μF, 25-volt ceramic disc capacitor
CR1	1N34A/1N90 germanium diode
CR2	1N4148 silicon diode
CR3	1N34A/1N90 germanium diode
CR4	1N4148 silicon diode
IC1	LM324 National Semiconductor
IC2	CD4066A RCA
IC3	LM324 National Semiconductor
IC4	LM386 National Semiconductor
M1	Ceramic or crystal microphone cartridge
SPKR	4-ohm miniature speaker
R1	1-megohm, ¼-watt, 10% composition resistor
R2	100K, ¼-watt, 10% composition resistor
R3	100K, ¼-watt, 10% composition resistor
R4	470K, ¼-watt, 10% composition resistor
R5	1-megohm, ¼-watt, 10% composition resistor
R6	1K, ¼-watt, 10% composition resistor
R7	820-ohm, ¼-watt, 10% composition resistor
R8	820-ohm, ¼-watt, 10% composition resistor
R9	820-ohm, ¼-watt, 10% composition resistor
R10	10K, ¼-watt, 10% composition resistor
R11	100-ohm, ¼-watt, 10% composition resistor
R12	47K, ¼-watt, 10% composition resistor
R13	2.2-megohm, ¼-watt, 10% composition resistor
R14	2.2-megohm, ¼-watt, 10% composition resistor
R15	10K, ¼-watt, 10% composition resistor
R16	47K, ¼-watt, 10% composition resistor
R17	1-megohm, ¼-watt, 10% composition resistor

R18	10K, ¼-watt, 10% composition resistor
R19	1.5K, ¼-watt, 10% composition resistor
R20	220K, ¼-watt, 10% composition resistor
R21	100K, ¼-watt, 10% composition resistor
R22	220K, ¼-watt, 10% composition resistor
R23	100K, ¼-watt, 10% composition resistor
R24	10K, ¼-watt, 10% composition resistor
R25	100K, ¼-watt, 10% composition resistor
R26	25K volume control potentiometer
S1	Dpst toggle switch

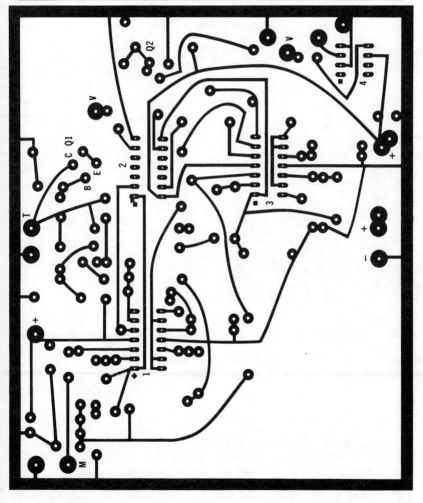

Fig. 9-4. Printed circuit layout shown full size (foil side).

Such use provides ease of service should it ever be necessary. When mounting the diodes be sure to follow the correct polarity as shown in Fig. 9-5. Do not interchange the silicon diodes (CR2 and CR4) with the germanium diodes (CR1 and CR3). The circuit will not work properly if this is done. After mounting the electrolytic capacitors check to see that they were installed with the correct polarity before soldering them in place. Fig. 9-5 indicates the polarity of the electrolytic capacitors with a + sign for the positive side of the capacitor.

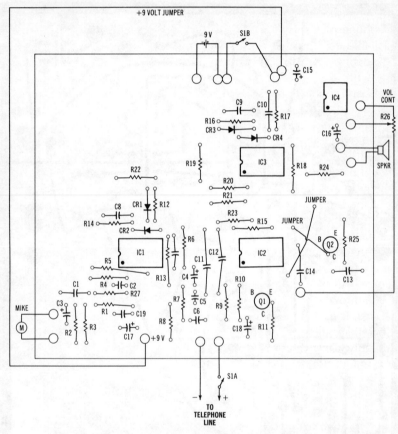

Fig. 9-5. Component layout (component side). Note three jumpers required as shown.

Be certain that the integrated circuits are inserted into the board in the correct direction. Pin 1 of each IC is indicated both on the foil layout and the component layout by a small dot next to pin 1. If an integrated circuit is incorrectly placed into the board, it could be permanently damaged when the power is turned on.

The printed circuit board will be connected to the components mounted in the cabinet with several wires. Use sufficient length of wire so that the board can be operated outside the cabinet for checkout and service. It is recommended that the connection to the microphone be made with shielded wire. You may use shielded wire for the other connections to the volume control and speaker if desired.

Since the Intercom Phone will be directly connected to the telephone line you should properly connect it by using a standard 4-prong telephone plug or modular plug for the connection. Such plugs are available from many electronics supply houses. Before you wire the connecting plug you will have to check the polarity of your telephone line with a dc voltmeter so that it agrees with the polarity of the telephone connections as shown in the schematic. This is necessary so that the circuit of Q1 operates properly when Intercom Phone is in use.

The size and style of the cabinet is left up to the builder. The only requirement is that the loudspeaker and microphone that are to be used should be mounted side by side facing the front of the unit, which should be covered by grille cloth so that it is transparent to the sound energy to and from the microphone and loudspeaker. Provision has been made on the printed circuit layout to mount the 9-volt battery on the board if so desired. If your cabinet is large enough, you can mount the battery at any convenient place inside the unit.

Intercom Phone has just two operating controls, volume control and off-on switch, which can be mounted at any desired place on the cabinet. Any style of switch may be used, as long as it is a double-pole single-throw switch. Before mounting the various parts inside the cabinet, keep in mind that you may want to add Dial Tone (Chapter 10) to provide calling capability to Intercom Phone. If this is done you will probably want to mount the touch tone pad on the top of the cabinet for convenient use.

PRELIMINARY CHECKOUT

Before attempting to use Intercom Phone it is recommended that it be given a preliminary checkout to determine that it is operating properly. To perform the following tests you will need an oscilloscope and audio oscillator.

Connect the output terminals of the audio oscillator to the telephone line connections of Intercom Phone as shown in Fig. 9-6. The resistance network shown is necessary so that the audio oscillator presents an impedance of about 820 ohms to the bridge circuit. Set the frequency of the audio oscillator to about 400 Hz and turn the output level to zero. Turn the power switch of Intercom Phone on and examine the signal level at pin 1 of IC1A with the oscilloscope as you speak into the unit with a normal speaking voice. IC1A should have an output signal level of about 1 volt peak to peak. Check the audio voltage level at pin 7 of IC1B with the oscilloscope. It should be the same as measured at pin 1, indicating that Intercom Phone is in send mode and switch IC2A is closed.

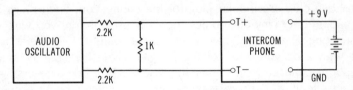

Fig. 9-6. Audio oscillator connected to intercom phone to simulate telephone line.

Set the oscilloscope vertical amplifier for dc operation and examine the voltage level at pin 7 of IC3B. This level should be about zero volts. Very slowly increase the output voltage level of the audio oscillator until the voltage at pin 7 of IC3B switches to about 9 volts. If you have the volume control of Intercom Phone set to about halfway you should hear the 400-Hz tone as IC3B switches to 9 volts. Check the level of the 400-Hz tone at the positive telephone line terminal of Intercom Phone which is necessary to cause IC3B to switch. This level should be about 0.1 to 0.2 volt peak to peak. Turn the output of the audio oscillator to zero. IC3B should switch back to zero volts in less than a second.

If Intercom Phone passes the foregoing tests you can assume it is operating properly and connect it to the telephone line for final checkout.

INSTALLATION AND FINAL CHECKOUT

Before connecting Intercom Phone to the telephone line check the polarity of the dc voltage appearing across the line with a dc voltmeter. This is very important since Intercom Phone will not work if you happen to connect it to the telephone line with the wrong polarity. As shown in the schematic diagram, the negative lead of the telephone line is connected to the common wire of the printed circuit board, and the positive lead of the telephone line is connected to the off-on switch. Use the standard 4-prong plug or modular plug for the connection to the telephone line.

You will not be able to dial telephone calls from Intercom Phone since it does not contain the necessary circuitry. (Chapter 10, Dial Tone, can be added to the unit to provide this capability if desired). To make a telephone call you will have to dial the number from your standard telephone and then turn on Intercom Phone after you have completed dialing. If you happen to have a push-button phone you can turn on Intercom Phone before you dial. Once the number has been dialed, turn on Intercom Phone and hang up the standard telephone. To operate the unit it will not be possible to have any other extensions on the line at the same time, since the low impedance of a standard telephone set will prevent Intercom Phone from operating properly.

To check the operation of Intercom Phone call up a friend. Explain what you are using and tell him or her that you can be heard only after the conversation stops at the other end. Also, there should not be any loud background noises at either end, such as a loud radio or tv, since this may prevent Intercom Phone from switching back and forth between send and receive.

You may leave Intercom Phone permanently connected to the telephone line, and it will not affect telephone performance in any way since it will be disconnected by the off-on switch when it is not in use. Anytime your telephone rings you can answer the call by turning on Intercom Phone, if you so desire.

Chapter 10

Dial Tone

Touch Tone dialing has been with us for a good number of years now, and anyone who has used it will agree that it is more convenient than the old-fashioned rotary dial. The only problem is that many telephone companies will charge an extra tariff each month to those customers who are using Touch Tone dialing. This can amount to a fair sum of money over a long period of time. There is an answer to this, however, through the development of a single integrated circuit chip which contains all of the frequency-synthesizing circuitry necessary to produce the required audio tones. It is a simple matter, therefore, to build this circuit and replace the rotary dial in an ordinary telephone with it. There is a wide variety of low cost telephones on the surplus market which can be converted to Touch Tone. You can also obtain the integrated circuit, 12-button keypad, and 1 MHz crystal from several electronics supply houses at reasonable cost. The circuit requires only a few extra components in addition to those mentioned above.

Dial Tone is small enough to be concealed within the housing of the K500 telephone (Fig. 10-1), and may fit into some of the new slim-line style telephones. When you have completed Dial Tone, you will have the equivalent of a telephone company Touch Tone telephone (Fig. 10-2). The circuit requires no external power source and is operated by the normal current supplied through the telephone lines. Since it has been designed to be used with the telephone system, it will operate properly and have no effect on telephone performance. The tone circuitry is active only during dialing and cannot be detected by the telephone company.

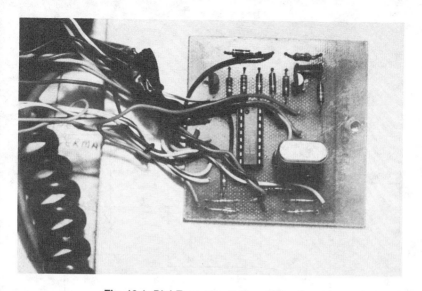

Fig. 10-1. Dial Tone circuit is very small.

Fig. 10-2. Appearance of a completed K500 telephone converted to Dial Tone.

ABOUT THE CIRCUIT

The complete schematic of Dial Tone is shown in Fig. 10-3. The heart of Dial Tone is IC1, the MC14410P integrated circuit that has been

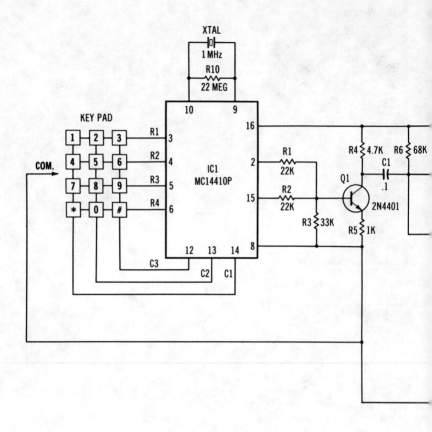

Fig. 10-3. Complete

developed by Motorola Semiconductor Products Inc. specifically for telephone use. This chip is capable of synthesizing a total of eight frequencies, two at a time, as required by the telephone system. (Dial Tone uses only seven frequencies with a 12-button keypad.) The circuit is crystal controlled, which results in accurate, stable frequencies that require no tuning adjustments. Refer to "Basic Telephone Principles" in Section I of this handbook for a discussion of the Touch Tone dialing system.

The internal circuitry of IC1 includes a crystal oscillator followed by two sets of divider chains, counters, decoders, and sine-wave generators that synthesize two tone frequencies in response to the selected push button. Built-in circuitry prevents erroneous tone generation in the event that two push buttons are pressed simultan-

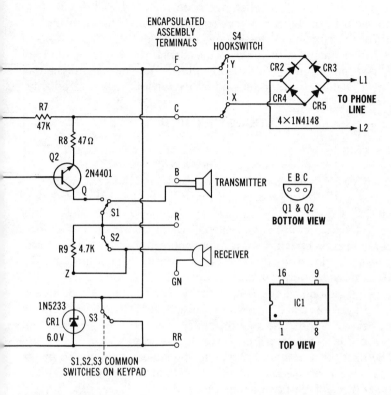

schematic of Dial Tone.

eously. Although the tones are generated by digital circuitry, they have low harmonic distortion and exceed minimum requirements for telephone network use.

The tone frequencies are divided into two groups, low band and high band. Pins 2 and 15 of IC1 are the output terminals, and drive Q1 through resistors R1 and R2. The mixed high band and low band frequencies are amplified by Q1 and fed to the telephone network through C1 and Q2. A portion of the mixed frequencies amplified by Q2 is fed to the receiver through R9.

Power to operate IC1, about 6 volts dc, is provided by the telephone network and zener diode CR1. When no button is pressed CR1 is shorted out by S3, one of the common switches of the keypad. This disables IC1 so that no tones can be generated. Whenever any button is pressed S3 opens. This causes 6 volts dc to be applied to IC1, and allows it to generate the tone frequencies.

A full-wave bridge rectifier has been included in the circuit to eliminate the possibility of a reverse polarity connection between the Dial Tone circuit and the telephone line. Diodes CR2, CR3, CR4, and CR5 ensure that the positive side of the telephone line is fed to pin 16 of IC1, and the negative side is fed to pin 7. Thus, it is permissible to connect Dial Tone to the telephone line without regard to polarity.

CONSTRUCTION

The parts list is given in Table 10-1. Dial Tone is constructed on a small printed circuit measuring about 2¼ × 2½ inches (6 cm × 6 cm), and will easily fit into the K500 style telephone. The small size of the printed circuit allows it to be mounted on the back of the keypad, or to the inside bottom of the telephone set. You may want to allow a little extra printed circuit board material when constructing the unit, depending upon how it is going to be mounted. Fig. 10-4 is a full size layout of the printed circuit board as seen from the copper side, and Fig. 10-5 is the view from the component side.

When mounting components to the printed circuit board be sure to check the polarity of the diodes before installation. Do not mix CR1 with the other four diodes. Use a socket for IC1. This permits an electrical test of the circuit before applying power to the chip.

The rotary dial of the K500 type telephone can be replaced by the keypad which has been mounted on a 4½-inch (11.43 cm) diameter circular piece of plastic or aluminum. A disc of this size can be easily epoxied into place in the plastic telephone housing. If you have a different style of telephone you may have to use a slightly different mounting procedure for the keypad.

Telephone style keypads are available from several electronics surplus supply houses. This circuit requires the standard telephone company style keypad which has a common switch assembly in addition to the 12 push buttons. This assembly consists of a spdt switch and two spst switches, normally closed. These three auxiliary switches are actuated whenever any push button is pressed. Be sure the keypad you are going to use closes two out of seven circuits, and is not the simplified single-pole single-throw style.

The telephone set wiring will have to be modified slightly when connecting the printed circuit. The following is a step by step procedure which will allow easy conversion. Refer to Figs. 10-3 and 10-6 which

Table 10-1. Dial Tone Parts List

Item	Description
C1	0.1-μF, 50-volt monolithic capacitor
R1	22K, ¼-watt, 10% composition resistor
R2	22K, ¼-watt, 10% composition resistor
R3	33K, ¼-watt, 10% composition resistor
R4	4.7K, ¼-watt, 10% composition resistor
R5	1K, ¼-watt, 10% composition resistor
R6	68K, ¼-watt, 10% composition resistor
R7	47K, ¼-watt, 10% composition resistor
R8	47-ohm, ¼-watt, 10% composition resistor
R9	4.7K, ¼-watt, 10% composition resistor
R10	22-megohm, ¼-watt, 10% composition resistor
CR1	1N5233 6.0-volt zener diode
CR2	1N4148 silicon diode
CR3	1N4148 silicon diode
CR4	1N4148 silicon diode
CR5	1N4148 silicon diode
IC1	Motorola MC14410P
Q1	2N4401 transistor
Q2	2N4401 transistor
Crystal	1 MHz
Keypad	12 button, 2 of 7 contact closure with auxiliary spdt and 2 spst switches

are schematic diagrams of the completed telephone and the telephone before conversion.

1. Remove the existing dial of the telephone by loosening one screw at each side of the dial assembly. Remove 4 dial wires from terminals R, GN, RR, and F of the telephone encapsulated assembly. Do not disturb any other wires which may be on these terminals, and tighten the screws that were loosened.

2. Identify the hookswitch wires which connect terminal L2 to C of the encapsulated assembly. There may be two hookswitch wires on terminal L2 as shown in Fig. 10-6.

3. Remove the hookswitch wires from terminal L2 and connect them to terminal X of the printed circuit board (Fig. 10-5). If you use a wire to make the connection, insulate the exposed splice with tape.

4. Connect a wire between terminal C of the printed circuit board and terminal C of the encapsulated assembly. Do not remove the existing hookswitch wire that is on terminal C of the assembly.

5. Identify the hookswitch wires which connect terminal L1 to terminal F of the encapsulated assembly. Remove the wire from

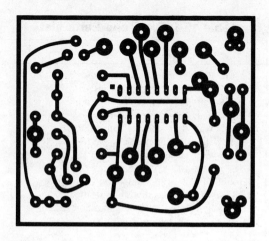

Fig. 10-4. Printed circuit layout shown full size (foil side).

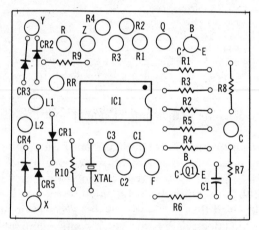

Fig. 10-5. Component layout (component side).

terminal L1 and connect it to terminal Y of the printed circuit board.

6. Connect terminal F of the printed circuit board to terminal F of the encapsulated assembly. Do not disturb the existing hookswitch wire on the assembly.

7. Connect terminal RR of the printed circuit board to terminal RR of the encapsulated assembly.

8. Remove a transmitter wire from terminal R of the encapsulated assembly, and connect it to the "break" terminal of the spdt switch on the keypad.

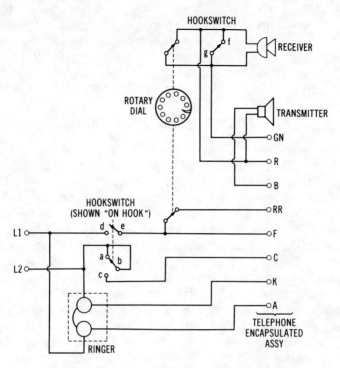

Fig. 10-6. K500 telephone before modification.

9. Remove a receiver wire from terminal R of the encapsulated assembly and connect it to terminal Z of the printed circuit board and "break" terminal of one of the spst switches on the keypad.

10. Connect together the "arm" terminal of the two common switches just wired, and also connect them to terminal R of the encapsulated assembly and terminal R of the printed circuit board.

11. Connect the "make" terminal of the spdt common switch to terminal Q of the printed circuit board.

12. Connect the last remaining spst common switch across terminals F and RR of the encapsulated assembly. Do not disturb any existing wires on these terminals.

13. Connect the keypad row and column switch common wire to terminal RR of the printed circuit board.

14. Connect the seven row and column wires of the keypad push-button switches to their respective terminals of the printed circuit board.

15. Connect a pair of wires to terminals L1 and L2 of the printed circuit board, and terminate them in a standard 4-prong or modular telephone connector. These wires will be the connections to the telephone line. This completes wiring of Dial Tone.

CHECKOUT AND OPERATION

Before attempting to use Dial Tone it is recommended that you make a voltage measurement on the printed circuit before inserting IC1 in its socket. This will prevent damage to the integrated circuit in the event of a possible wiring error. To make this measurement, connect the L1 and L2 wires coming from the printed circuit board to the telephone line. You do not have to worry about the polarity of the line. Lift the telephone receiver off the hook, press any push button on the keypad and measure the voltage between pins 16 and 8 of the integrated circuit socket. This voltage should be within the range of 5.7 to 6.3 volts, with pin 16 being positive with respect to pin 8. If the correct voltage reading is obtained disconnect the circuit from the telephone line before inserting IC1 in its socket. If you obtain an incorrect voltage reading troubleshoot the circuit before proceeding further. When inserting IC1, be sure to orient it in the correct direction as illustrated in Fig. 10-5.

Once IC1 has been inserted in its socket Dial Tone is ready to be used. You then will be able to dial telephone numbers by means of the pushbuttons of the keypad.

Chapter 11

Auto-Dial

Some telephone exchanges, especially the older nonelectronic types, cannot make use of the Touch Tone system which permits dialing numbers by means of push buttons. This problem has been solved through the use of two specialized digital integrated circuits manufactured by Motorola Semiconductor Products, Inc. These are the MC14419 and MC14409 low-power complementary MOS chips which are designed for use in telephone pulse dialing applications. The MC14419 is a keypad-to-binary code converter which accepts data on seven input terminals and converts it to binary form which can be used by its companion device, the MC14409. The combination of the two chips will convert the selected push-button number value to a series of output pulses which can be utilized by the telephone network to operate the stepping relays and dial a telephone number. The entire circuit operates on extremely low power which is derived from the telephone line itself, and has memory capability to memorize the last number dialed. Thus, this circuit has a distinct advantage over the standard Touch Tone system in that you can redial a number as many times as you want simply by pushing one button. This will be advantageous when you get a busy signal and need to call again a few minutes later.

Auto-Dial can be used with any telephone by wiring the circuit to the telephone according to the specified procedure. Once the circuit is connected to the telephone the normal rotary dial will be inoperative, since it no longer will be necessary.

ABOUT THE CIRCUIT

When the telephone is not in use the hookswitch opens the connections between the telephone line and terminals F and C of the encapsulated circuitry, just as happens in a standard unmodified telephone. The built-in regulated power supply is active, however, drawing a very small current from the telephone line. When the telephone handset is lifted, terminals b and c of the hookswitch close, providing forward bias to Q2 through R5. This causes Q2 to draw sufficient current from the telephone line to cause the switching equipment at the central office to deliver a dial tone, and wait for the dial pulses to follow.

Refer to Fig. 11-1 which is a block diagram, and to Fig. 11-2, the timing diagram of the telephone dialer system as used by Auto-Dial. The twelve-button keypad has been designed to close two circuits when any one button is pressed. These circuits are divided into two groups, rows and columns. If you look at the graphic representation of the keypad you will note that each key is tied to a row and a column wire. The keypad common is connected to the system "ground," V_{ss}, so that one of the row wires and one of the column wires will be grounded when any key is pressed. This information is converted to binary form by IC1 and presented to IC2 on four wires, D1, D2, D3, and D4. A fifth wire,

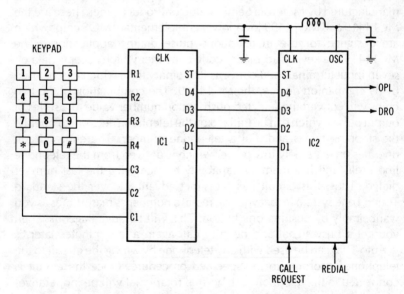

Fig. 11-1. Block diagram of Auto-Dial circuit.

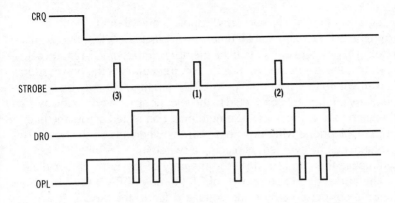

Fig. 11-2. Timing diagram showing first three digits (312) of typical phone number.

Strobe, is used to signify that the data on the four lines is valid, and enters the 4-bit binary number into the internal First In, First Out memory of IC2. The first strobe pulse occurring after the call request line is taken low clears the memory of any previous telephone number and enters the first digit of the new number. The circuit has a total capacity of sixteen digits, and will ignore any digits in excess of that amount. If the Redial line is taken low instead of pressing one of the numbered push buttons, the circuit will automatically outpulse the digits stored in the memory. Outpulsing of the telephone number is provided by the OPL output of IC2, which cuts off Q2 and opens the telephone line just as the rotary dial switch would do in a normal telephone. The Dial Rotating Output (DRO) terminal of IC2 goes high each time a digit is being sent by IC2. This cuts off Q1 and prevents the dial pulses from being heard in the receiver. The operating speed of the circuit is determined by the frequency of the internal clock oscillator. L1, C1, and C2 set the frequency of the oscillator to about 16 kHz so that the outpulsing rate of the circuit is about 10 pulses per second. If desired, the outpulsing rate may be increased to 20 pulses per second by reducing the value of C1 and C2 by a factor of 4. This range is within the operating capability of the telephone network.

Power to operate the circuit is supplied by the telephone network, so that no external power source is required. This is possible since complementary MOS integrated circuits operate on microwatts of power. R1, CR1 and C3 form a regulated power supply that produces a low-voltage power source to operate IC1 and IC2. Zener diode CR1 prevents the voltage from exceeding 6 volts, the maximum rating of the integrated circuits. The regulated supply draws about 0.2 milliampere from the telephone line at all times, even when the telephone is on the hook. This is necessary so that the memory circuits of IC2 remain active and store the last number dialed. The current drain of the power supply is well within the maximum permissible amount that may be taken by auxiliary equipment connected to the telephone network, but may alert the telephone company that something is connected across the line. A normal telephone will draw no dc current when it is on the hook. This problem can be avoided by placing a switch in series with R1, but this will cause the circuit to lose the memory of the last number dialed.

The bridge circuit composed of CR2 through CR5 is used to ensure that the correct polarity of dc voltage is fed to the circuit. It will not matter which way the circuit is connected to the telephone line since the bridge rectifier will always feed the positive side of the line to V_{dd} and the negative side of the line to V_{ss}. This avoids the problem of damaged integrated circuits due to accidental misconnections.

CONSTRUCTION

The parts list is given in Table 11-1. Auto-Dial may be constructed in two ways. You can assemble the entire circuit and push-button assembly in the confines of a standard K500 or similar telephone, or you can use a separate case for the circuit and connect the two units together

Fig. 11-3. Auto-Dial housed in separate case.

Table 11-1. Auto-Dial Parts List

Item	Description
C1	0.047-μF, 100-volt tubular capacitor
C2	0.047-μF, 100-volt tubular capacitor
C3	100-μF, 10-volt electrolytic capacitor
R1	220K, ¼-watt, 10% composition resistor
R2	68K, ¼-watt, 10% composition resistor
R3	68K, ¼-watt, 10% composition resistor
R4	470K, ¼-watt, 10% composition resistor
R5	100K, ¼-watt, 10% composition resistor
R6	1 megohm, ¼-watt, 10% composition resistor
R7	100K, ¼-watt, 10% composition resistor
L1	3.3-millihenry inductor
Q1	Motorola MPS-L01 transistor
Q2	Motorola MPS-L01 transistor
IC1	Motorola MC14419 IC
IC2	Motorola MC14409 IC
CR1	1N4734 zener diode
CR2	1N4148 silicon diode
CR3	1N4148 silicon diode
CR4	1N4148 silicon diode
CR5	1N4148 silicon diode
CR6	1N4148 silicon diode
CR7	1N4148 silicon diode
CR8	1N4148 silicon diode
CR9	1N4148 silicon diode
Keypad	Telephone type, closes two out of seven circuits

with a cable, as shown in the photograph of Fig. 11-3. The printed circuit layout is shown full size in Fig. 11-4, and the component layout is shown in Fig. 11-5. Since the integrated circuits used in Auto-Dial are not low cost, it is strongly recommended that IC sockets be used. This will permit you to fully check the circuit out with power applied before you insert the ICs into the circuit. This will prevent damage to an IC in the event of a miswire or short circuit. The printed circuit has 22 connections to the push-button keypad and telephone, and turret lugs for these connections are recommended if they are available to you. Be sure to check the polarity of the diodes, transistors, and integrated circuits before placing them in the circuit. This will avoid the possibility of damaged components and eliminate a great deal of troubleshooting when the circuit is placed in operation. The integrated circuits will probably be supplied to you mounted in conductive foam. Do not remove them from this foam or place them in the circuit until instructed to do so during checkout of the circuit. They can be easily damaged by static electricity charges.

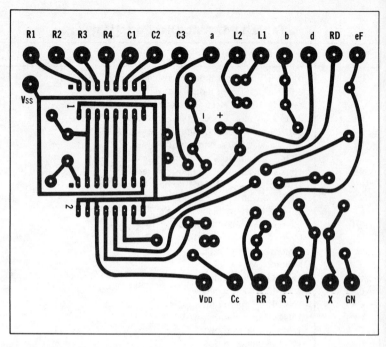

Fig. 11-4. Printed circuit layout shown full size (foil side).

The printed circuit assembly is connected to the telephone with several wires, and some of the connections inside the telephone must be changed to operate with Auto-Dial. The following procedure is based on a K500 telephone and will help you in performing the conversion. Other types of telephones may be used, but you will have to identify the proper terminals within the telephone if they are different from the K500 telephone. Fig. 11-6 shows the K500 telephone before modification, and Fig. 11-7 shows the entire circuit including printed circuit and telephone. Fig. 11-8 illustrates the terminal arrangement of the K500 encapsulated assembly. Refer to these diagrams when performing the conversion.

1. Remove the telephone dial by loosening two screws and lifting the dial out of its bracket. Disconnect two wires that connect the dial to terminals GN and R on the encapsulated assembly.
2. Remove the two remaining wires on terminal GN which go to the receiver and hookswitch. Solder these wires together and connect them to terminal X on the printed circuit board.
3. Connect the GN terminal on the encapsulated assembly to terminal GN on the printed circuit board.

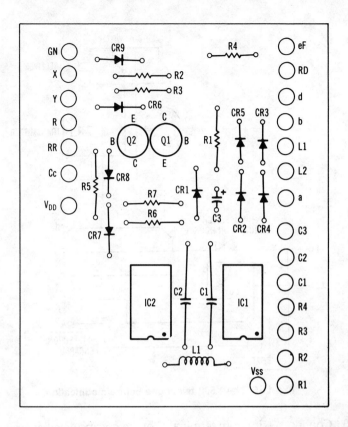

Fig. 11-5. Component layout (component side).

4. Remove the remaining receiver wire and hookswitch wire G from terminal R of the encapsulated assembly. Leave the transmitter wire connected to R. Solder the receiver wire and hookswitch wire G together and connect them to terminal Y of the printed circuit board.

5. Connect R of the encapsulated assembly to R of the printed circuit board.

6. Remove the dial switch wire from terminal RR of the encapsulated assembly. Connect terminal RR of the encapsulated assembly to RR of the printed circuit board.

7. Remove the dial switch wire from terminal F of the encapsulated assembly, but leave the hookswitch wire connected to F. Connect terminal F of the encapsulated assembly to terminal F of the printed circuit board.

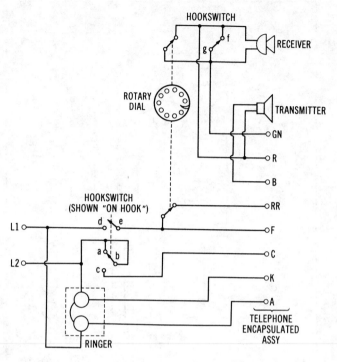

Fig. 11-6. The K500 telephone before modification.

8. Connect a wire from terminal C of the encapsulated assembly to terminal Cc of the printed circuit board. This will require soldering to the terminal on the encapsulated assembly. Do not disturb the hook switch wire that is already connected to terminal C of the assembly.

9. Connect L1 on the encapsulated assembly to terminal d of the printed circuit board. Note that this wire is also connected to wire d of the hookswitch.

10. Locate the remaining two hookswitch wires, a and b. Identify these wires by using an ohmmeter and checking the switch action of the single-pole double-throw circuit composed of wires a, b, and c. Connect wire a to terminal a of the printed circuit board, and connect wire b to terminal b of the printed circuit board. You may use the unused terminal L2 of the encapsulated assembly as a convenient connection to wire b.

11. Connect a pair of wires to terminals L1 and L2 of the printed circuit board. These wires will be connected to the telephone

line. Use a standard 4-prong telephone or modular connector as a termination for these wires.

12. Connect the keypad switches to terminals C1, C2, C3, R1, R2, R3, and R4 of the printed circuit board as shown in the schematic diagram. Connect the keypad common wire to terminal V_{ss} of the printed circuit board.

13. Wire the Redial switch as shown in the schematic. Terminal RD of the printed circuit board should be connected to V_{dd} when the Redial switch is in the released position, and connected to V_{ss} when pressed. This completes wiring of the printed circuit board and telephone.

CHECKOUT AND TEST

Checkout of the circuit is composed of two parts, ohmmeter checks and voltage checks. It is best to perform these checks before applying power to the integrated circuits to avoid any damage to them in the event of any possible miswiring.

Connect the common side of the ohmmeter to terminal V_{ss} of the printed circuit board and check the operation of each push button on the keyboard. Each button, when pressed, should indicate a connection of two wires to V_{ss} as shown in the schematic diagram. For example, pressing button number 1 should short pin 1 and pin 5 of IC1 to V_{ss}. Make this check on all push buttons. Check the operation of the hookswitch wires a, b, c, d, and e. Operate the hookswitch and ascertain that the switch works according to the circuit diagram. Check the operation of the Redial switch with the ohmmeter. Once these checks are completed you proceed to the next test.

Do not insert the integrated circuits into the printed circuit board until the following voltage checks have been made. Use a vtvm or similar high-impedance voltmeter for voltage measurements.

Power can be applied to the circuit in either of two ways. You can connect the circuit directly to the telephone line, or you can use a 48-volt dc power source which has a 1000-ohm, 2-watt resistor connected in series with the output. This simulates the source impedance of the telephone line. You can connect the L1 and L2 terminals of the printed circuit board to the power source without regard to polarity, since the full wave diode bridge in the circuit will direct the positive and negative lines of the power source in the correct direction.

Connect the common lead of the voltmeter to the V_{ss} terminal of the printed circuit board, and measure the voltage at the V_{dd} terminal. This

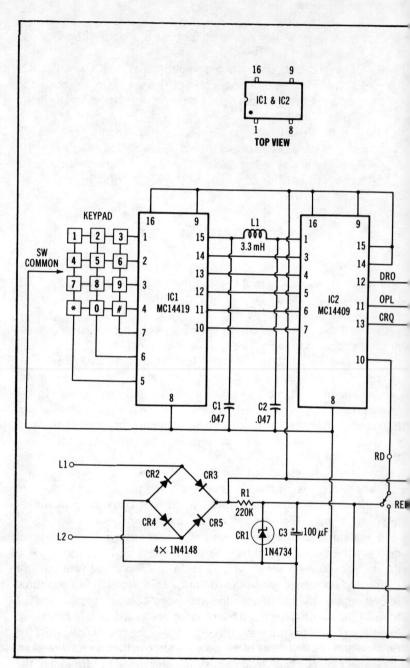

Fig. 11-7. Complete schematic diagram

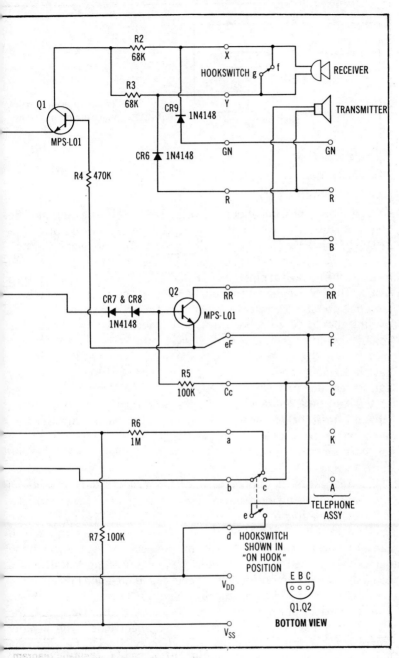

of Auto-Dial and telephone.

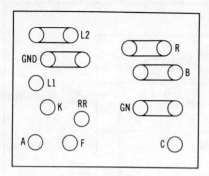

Fig. 11-8. Terminal arrangement of K500 encapsulated assembly.

voltage should measure about +5 volts. This will be the output of the built-in regulated power supply. If the voltage exceeds 6 volts, or is negative, check the circuit and components before proceeding further.

Leave the common lead of the voltmeter connected to V_{ss} and measure the voltage at pins 9 and 16 of IC1, and pins 9, 14, 15, and 16 of IC2. The voltage reading should be the same as measured at the V_{dd} terminal, about +5 volts. Measure the voltage at pin 8 of each IC. The reading should be zero. Measure the voltage at pin 10 of IC2. The voltage should be zero, and when you press the Redial pushbutton, the voltage should read about +5 volts. Once these voltage measurements have been made, and are correct, the circuit can be disconnected from the power source.

With the power removed from the circuit you can insert the integrated circuits in their sockets. Be very careful to position each IC in the correct direction by noting which side of the IC is pin 1 and referring to Fig. 11-5 for the correct placement. Also, be careful to place each IC in the correct socket.

You are now ready to operate Auto-Dial. Connect the circuit to the telephone line using the standard 4-prong telephone plug or modular connector. Lift up the telephone handset. You should hear a dial tone. Dial any telephone number by pushing the appropriate buttons. You can terminate the call at any time by hanging up, and the last telephone number you dialed will be stored in the memory of the Auto-Dial. To call that number, simply lift up the handset and press the Redial button once. The circuit will automatically outpulse all digits of the number. This can be done any number of times, but if you dial a new number the memory will retain only the last telephone number dialed. The memory will be active as long as the circuit remains connected to the telephone line, even though the telephone is inactive.

Chapter 12

Ring-A-Thing*

Did you ever miss an important telephone call because you were out of range of sound of the telephone bell? It doesn't have to happen again if you build and install this inexpensive remote telephone bell (Fig. 12-1). It is battery operated and can be located anywhere inside or outside your home. Since it is self powered it requires virtually no energy from the telephone line. The input impedance of the circuit, as seen by the telephone line, is almost 100,000 ohms and the input resistance is infinite. When connected across the telephone line it is undetectable and has no effect on telephone performance.

The circuit derives its power from four rechargeable NiCad cells connected in series which provide 4.8 volts to drive an ordinary doorbell. Since the power demand on these cells occurs only when the telephone rings, the battery will operate the bell over 1000 times on one charge. This should last several months, depending upon how many calls you receive. A built-in battery charger is included in the circuit so that the cells may be conveniently charged from the ac power line at any time. Full recharge takes 14 hours, but the charger may be left in operation indefinitely, if desired, with no damage to the cells due to overcharge. This is possible since the charging circuit has been designed to deliver a limited current to the cells. The NiCad cells used in this circuit are size C, but other sizes may be used if desired.

---

*Reprinted with permission from *Elementary Electronics Magazine*

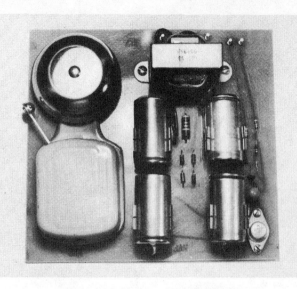

Fig. 12-1. Photo of Ring-A-Thing.

HOW IT WORKS

Refer to Fig. 12-2. When the telephone rings, a 20-Hz ac voltage of about 90 volts rms is impressed across the telephone line. The series circuit composed of R1, R2, R3, C1, and C2 is connected across the line to provide isolation and act as a voltage divider for Q1. C1 and C2 provide dc isolation, since the line normally has a dc voltage of about 48 volts across it when the telephone is not in use. Q1 responds to the 20-Hz ringing signal by conducting current during each positive half cycle applied to its base. CR1 prevents Q1 from being reverse biased during the negative half of the ringing signal. The emitter current of Q1 is applied to the base of Q2 causing it to saturate and act as a switch. This applies full battery voltage to the bell, causing it to ring. The voltage applied to the bell is essentially a 20-Hz square wave which produces a slightly different sound than that produced by pure dc. CR2 and C3 protect Q2 from any reverse voltage spikes produced by the collapsing magnetic field of the bell.

The battery charger circuit is composed of T1, a 4-diode bridge rectifier, and R5. T1 provides isolation from the ac power line while reducing the voltage to about 6 volts rms. The output of the bridge rectifier, a pulsating dc of about 9 volts peak, is applied to the four cells through a current-limiting resistor, R5. This type of circuit is

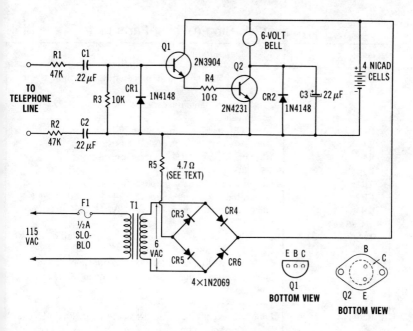

Fig. 12-2. Schematic diagram of Ring-A-Thing.

recommended for NiCad cells, and provides essentially a constant charge current regardless of the state of charge of the battery or power line voltage. By limiting the current to not more than 1/10 of the ampere hour rating of the cells, the charger may be operated for any length of time without damage to the battery due to overcharge. When the cells attain full charge, the gases produced within the cell are recombined chemically, thus preserving the electrolyte.

CONSTRUCTION

The parts list is given in Table 12-1. The entire circuit is built on a 6½ inch × 7 inch (16½ cm × 18 cm) printed circuit board. The foil layout is shown half size in Fig. 12-3 and the component layout is shown in Fig. 12-4. The cells are securely mounted to the printed circuit board using steel clips. This method of assembly is recommended since it would not be good practice to rely on the connecting wires of the cells to hold them in place. If the cells you are using do not have solder tabs, the wires can be soldered directly to the positive and negative metal parts of the cell. In this case do not use excessive heat when soldering so that the cells do

Table 12-1. Ring-A-Thing Parts List

Item	Description
C1, C2	0.22-μF, 250-volt tubular or ceramic capacitor, Radio Shack 272-1070 or equivalent
C3	22-μF, 16-volt tantalum capacitor, Radio Shack 272-1412 or equivalent
R1, R2	47K, ¼-watt, 10% composition resistor
R3	10K, ¼-watt, 10% composition resistor
R4	10-ohm, ¼-watt, 10% composition resistor
R5	4.7-ohm, 1-watt, 10% composition resistor (see text)
CR1	Silicon diode, general purpose 1N4148 or equivalent
CR2	Silicon diode, general purpose 1N4148 or equivalent
CR3, CR4, CR5, CR6	Silicon diode, 0.5-amp, 100-volt PIV, 1N2069 or equivalent
T1	6.3-volt, 1-ampere filament transformer, Radio Shack 273-050 or equivalent
Q1	Silicon transistor, npn, 2N3904 or equivalent
Q2	Silicon power transistor, npn, 5-ampere 2N4231 or equivalent
F1	½-ampere slow-blow fuse
Battery	Four NiCad cells size C, 1.2-ampere hour, Radio Shack 23-124 or equivalent
Battery Clips	Augat 6020 or equivalent

not become damaged. When mounting the cells be sure to follow the exact polarity as shown in Fig. 12-4.

The printed circuit layout for the bell connections may be changed to accommodate the type of bell you are going to use. Be sure to locate the mounting holes for the bell before laying out the printed circuit to avoid a conflict between the copper foil and mounting screws. Do not solder R5 into the printed circuit until instructed to do so later in the test procedure. Q2 is mounted directly to the printed circuit board without a heat sink. None is required since the transistor operates as a switch, resulting in almost no power dissipation.

Use a 4-prong or modular connector to connect Ring-A-Thing to the telephone line. It is not necessary to observe polarity of the dc voltage appearing across the line, since the circuit is ac coupled to the line. If Ring-A-Thing is placed at a location which is not convenient to an ac power line, it will be a simple matter to disconnect it from the telephone line temporarily when the battery needs recharging.

A word of caution about NiCad cells. It is recommended that the cells be handled in a discharged state. NiCad cells are capable of delivering very large short circuit currents (50 amperes or more) even when only partially charged. Once the cells are mounted and wired to the printed circuit any accidental short circuit between the cells or other components on the board may cause very large current flow. Such currents can easily burn out printed circuit wiring.

CHECKING THE CIRCUIT

After the unit is assembled and wired, it is recommended that the charger current be measured so that the proper value of R5 can be

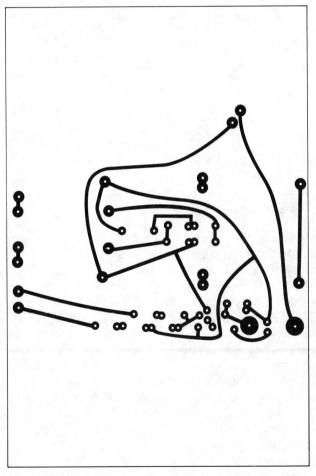

Fig. 12-3. Printed circuit layout shown half-size (foil side).

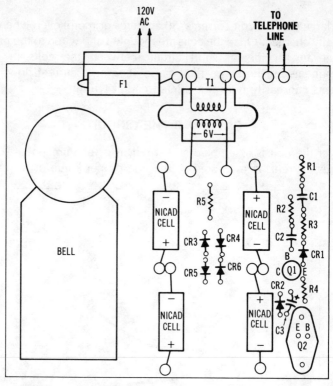

Fig. 12-4. Component layout shown from top of board.

placed in the circuit. Since different 6.3-volt filament transformers can vary considerably in output voltage and internal impedance, the current delivered by the charger should be checked and adjusted if necessary. The best method of measuring charging current is to insert a 0-1 ampere dc ammeter in series with R5. An alternate method is to measure the resistance of R5, connect it into the circuit, and measure the dc voltage across it. Current can then be calculated by dividing the voltage measurement by the resistance.

The recommended charge current for the C cells specified in this parts list is 120 milliamperes. This will charge the battery in 14 hours, and there would be no danger of overcharge if the line power was connected for several days. If you prefer to leave the charger permanently connected to the ac power line the charge current should be reduced by a factor of 3, to 40 milliamperes. This will keep the cells at 100% charge without any danger of cell damage. If NiCad cells of other capacities are used the charge current should be set up to 1/10 of the ampere hour rating of the cells, or to 1/30 the rating if the charger is to be operated permanently. For example, if size D cells with a 3.5-ampere hour rating were used, the charge current would be set to 350 milliamperes or 115 milliamperes. The value of R5 should be changed, if necessary, to provide the desired current. Note that it is not necessary to set the current to exactly 1/10 of the rating of the cells. Less current can be used with the disadvantage being that it would take longer than 14 hours to fully charge the battery. Do not use a larger current than specified. To do so will damage the cells on overcharge.

INSTALLATION

The unit is connected across the telephone line as shown in the schematic. The only exception to this will be for two party telephones. In this case the telephone ringing signal is impressed between one of the telephone lines and ground. (The other party's ringing signal is impressed across the other line and ground). You will have to experiment to determine which of the two lines has your ringing signal. If you inadvertently connect the circuit to the wrong line, you will be answering the other party's calls! For the ground connection you may use any convenient ground point such as a BX ground.

Before installing the unit you should operate the charger at least 14 hours to fully charge the battery, unless you plan to leave the charger connected permanently to the power line. With a fully charged battery the unit will operate several months before a recharge is necessary. The power demand of the charger is about 2 watts, and will have little effect on your electric bill if left operating. In this case the unit would need no further attention.

Chapter 13

Telephone Trigger*

How would you like to be able to energize any electrical device in your home, from anywhere in the country, and do it without paying for a phone call? You can do it, and it's perfectly legal. The remote control described here is a simple digital circuit which responds to the sound of the telephone bell (Fig. 13-1). There is no need to make any hard wire connections to the telephone line, and it is this feature which permits you to build and use this device without any permission from Ma Bell. You can't even be charged any tariff for using it. In these days of high energy costs it would be economical to shut the heat or air conditioning off before you left the house, and turn it on about an hour before you arrive home.

The remote-control circuit is protected against accidental operation through normal telephone calls by means of an automatic reset feature which cancels out the effects of any telephone calls made by others. This is accomplished by incorporating a time-delay circuit which allows the circuit to receive a valid code for a period of ninety seconds. When the ninety second period is completed, the circuit resets itself and waits for the next telephone call.

The proper code which activates the circuit consists of 2 rings of the telephone, a 25- to 40-second delay, and 2 more rings. If, and only if, this code is received before the 90-second delay is terminated will the device be activated. Any other combination of rings will not operate the circuit. Since it is very unlikely that anyone would ring your telephone

---

*Reprinted with permission from *Elementary Electronics Magazine*

Fig. 13-1. Photo of Telephone Trigger.

with such a sequence, accidental operation is virtually eliminated. Included in the circuit is a group of LED indicators which monitors the control pulses and indicates the status of the circuit at all times.

HOW IT WORKS

The best way to understand circuit operation is to refer to Fig. 13-2 which illustrates several pertinent waveshapes in the circuit. The schematic diagram is given in Fig. 13-3. A crystal or ceramic microphone is used as the sensing element which detects the sound of the telephone bell. The output of the microphone is fed to the negative input of a comparator, IC1. The positive input of the comparator is set to a positive dc voltage by means of a potentiometer which acts as a sensitivity control. This forces the output of IC1 to +5 volts. During periods of silence there is insufficient output from the microphone to exceed the voltage setting of the sensitivity control, and the output of the comparator remains at 5 volts. When the telephone rings, the increase in sound energy causes the output of the microphone to exceed the setting of the sensitivity control. The output of the comparator oscillates between zero and 5 volts as the bell continues to ring. This is shown in Fig. 13-2, waveform A.

The output of IC1 is fed to the trigger input terminals of IC2 and IC3, pin 2. Each of these ICs is a 555 timer connected to operate as a

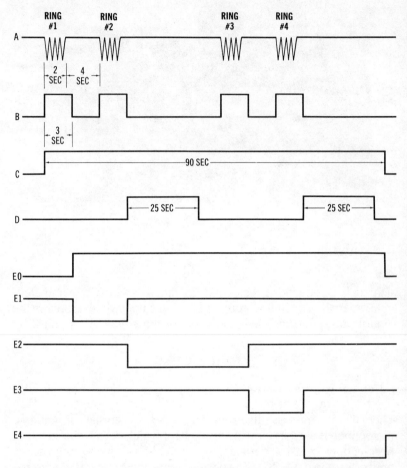

Fig. 13-2. Timing diagram showing logic levels at several points in the circuit.

monostable or one-shot multivibrator. IC2 produces an output pulse of about 3 seconds duration, and IC3 produces an output pulse of about 90 seconds duration. These output pulses are shown in Fig. 13-2 as waveforms B and C.

The purpose of IC2 is to convert the rapidly oscillating output of IC1 to a rectangular pulse of known duration, about 3 seconds. Note that the pulse time of IC2 is greater than that of one ring, but stops before the start of the next ring. IC2 provides an accurate waveform which can be counted by IC4.

The output of IC2 is inverted by IC7C, which in turn drives LED "R."

This provides a visual indication of the circuit response to the sound of the telephone.

IC3 is used as the control of the binary counter IC4. When the circuit is in a standby condition, the outpt of IC3 is at a logic level of zero. This is fed to IC4 reset terminals 2 and 3 and forces IC4 to be set to a count of zero. When the first telephone ring is received, the output of IC3 goes to a logic one state, thus allowing IC4 to count for a period of 90 seconds.

The output of IC2 feeds the clock input, pin 14, of IC4 which clocks on the trailing or falling edge of the pulses. IC5 is a 4-bit decoder which provides a zero logic level at any one of its output terminals as determined by the binary information fed from IC4 to its input terminals. The output of IC5 is used to drive a set of LEDs and also to control a 25-second timer, IC8. When IC4 reaches a count of 2, the 25-second timer is activated. This is shown in Fig. 13-2 as waveform D. If IC4 receives a third clock pulse from IC2 during this interval, it is reset to zero. Once IC8 returns to its normal state after the period of 25 seconds, IC4 is ready to receive additional clock pulses without being reset to zero.

When IC4 reaches a count of 4, the 25-second timer is reactivated, as shown in waveform D. The output of IC5, pin 5, is prevented from reaching Q1 base until the 25-second period is over. Should IC4 receive any more input pulses Q1 would not be activated at the end of the 25-second period, since the counter would then be at a count of 5 or more.

The outputs of IC5 are shown in Fig. 13-2 as waveforms E. When IC5 is fed a binary number from 0 to 9, the corresponding output terminal assumes a logic level of zero. All other output terminals remain at a logic level of one. By connecting LEDs to outputs 0, 1, 2, 3, and 4, (pins 1, 2, 3, 4, and 5) the status of binary counter IC4 is visually indicated.

The coil of a 12-volt relay is connected in the collector circuit of Q1 so that it is activated whenever Q1 conducts current in response to the zero logic level of IC5, pin 5. Since the output of IC5 is not permanent, one section of the relay contacts is connected to the coil circuit so that the relay latches and remains activated even though IC4 is reset to zero at the end of the 90-second pulse time of IC3. The other set of contacts of the relay is used as a single pole switch to operate the desired equipment.

Power to operate the circuit is obtained through a 12-volt transformer feeding a full-wave bridge rectifier. The output of the rectifier is fed to IC9 which is a fixed 5-volt regulator. The entire circuit, with the exception of the relay, operates on the 5-volt output of IC9. The coil circuit of the relay is returned to the unregulated 12-volt output of the bridge rectifier.

CONSTRUCTION

The parts list is given in Table 13-1. The entire circuit is constructed on a printed circuit board, measuring about 8½ inches × 4¼ inches (21 cm × 12 cm). The crystal microphone cartridge can be mounted directly on the board with adhesive, but also may be connected by means of a shielded wire to some remote location near the telephone bell. The foil pattern of the printed circuit board is shown in Fig. 13-4 in half-size, and the component layout is shown in Fig. 13-5. Although this

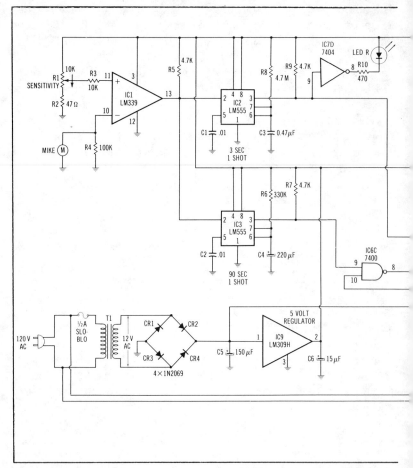

Fig. 13-3. Telephone Trigger

is a relatively simple digital circuit, it is recommended that sockets be used for the integrated circuits. The initial checkout of the circuit will be simplified if IC3 can be temporarily removed from the circuit. In the event that the unit ever requires service you will find that it is well worth the added cost of sockets. It is extremely difficult to remove a multipin IC which has been soldered directly into a printed circuit board without destroying the IC or printed wiring. When mounting the electrolytic capacitors, diodes, and LED indicators, be sure to observe the correct polarity as shown in the schematic and component layout.

Mount and solder all parts to the printed circuit board as shown in Fig. 13-5. After this is done you will be able to locate and insert the proper

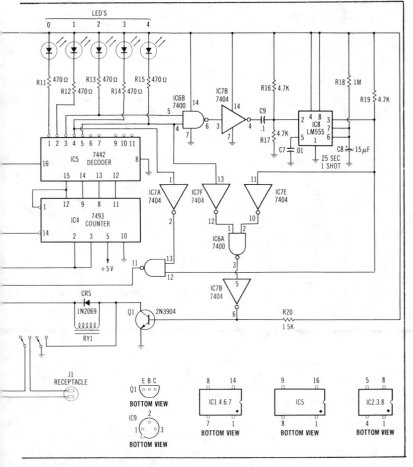

schematic diagram.

Table 13-1. Telephone Trigger Parts List

Item	Description
R1	10K potentiometer
R2	47-ohm, ¼-watt, 10% composition resistor
R3	10K, ¼-watt, 10% composition resistor
R4	100K, ¼-watt, 10% composition resistor
R5	4.7K, ¼-watt, 10% composition resistor
R6	330K, ¼-watt, 10% composition resistor
R7	4.7K, ¼-watt, 10% composition resistor
R8	4.7-megohm, ¼-watt, 10% composition resistor
R9	4.7K, ¼-watt, 10% composition resistor
R10	470-ohm, ¼-watt, 10% composition resistor
R11	470-ohm, ¼-watt, 10% composition resistor
R12	470-ohm, ¼-watt, 10% composition resistor
R13	470-ohm, ¼-watt, 10% composition resistor
R14	470-ohm, ¼-watt, 10% composition resistor
R15	470-ohm, ¼-watt, 10% composition resistor
R16	4.7K, ¼-watt, 10% composition resistor
R17	4.7K, ¼-watt, 10% composition resistor
R18	1-megohm, ¼-watt, 10% composition resistor
R19	4.7K, ¼-watt, 10% composition resistor
R20	1.5K, ¼-watt, 10% composition resistor
CR1	1N2069 silicon diode
CR2	1N2069 silicon diode
CR3	1N2069 silicon diode
CR4	1N2069 silicon diode
CR5	1N2069 silicon diode
C1	.01-μF, 25-volt, disc capacitor
C2	.01-μF, 25-volt disc capacitor
C3	.47-μF, 25-volt ceramic capacitor
C4	220-μF, 10-volt tantalum capacitor
C5	.150-μF, 15-volt tantalum capacitor
C6	15-μF, 10-volt tantalum capacitor
C7	.01-μF, 25-volt disc capacitor
C8	15-μF, 10-volt tantalum capacitor
C9	.1-μF, 25-volt ceramic capacitor
IC1	LM339 quad comparator
IC2	LM555 timer
IC3	LM555 timer
IC4	SN7493N binary counter
IC5	SN7442N decoder
IC6	SN7400N quad NAND gate
IC7	SN7404N hex inverter
IC8	LM555 timer
IC9	LM309H 5-volt regulator
F1	½ ampere slow blow fuse
Q1	2N3904 npn silicon transistor or equivalent

T1	6-volt transformer, Radio Shack 273-1385 or equivalent
LEDs	Radio Shack 276-1622 or equivalent
RY1	Relay Radio Shack 275-206 or 275-208 or equivalent (see text)
MIC	Microphone, Radio Shack 270-095 or equivalent
Misc.	Line cord, receptacle, IC and transistor sockets, hardware, wire and solder.

jumpers into the printed circuit board by referring to the schematic diagram.

Be sure to use a relay which is capable of carrying the current of the appliance which is to be turned on. The parts list shows a choice of two relays. Part number 275-206 has a current rating of 10 amperes. You will have to use the higher current relay if you plan to operate a high-current appliance such as an air conditioner or coffee maker. If the remote control is to be used to operate a heating system, the relay contacts can be connected into the thermostat circuit. Such a connection permits the use of the lighter duty relay. The relay coil driver transistor, Q1, can safely carry up to 150 milliamperes to drive the relay coil.

A receptacle for plug-in appliances is mounted on the circuit board and is wired directly to the line cord and relay as shown in the schematic. Be sure to use a line cord and wire which will safely carry the desired current. For 10 ampere operation use at least a No. 16 gauge wire.

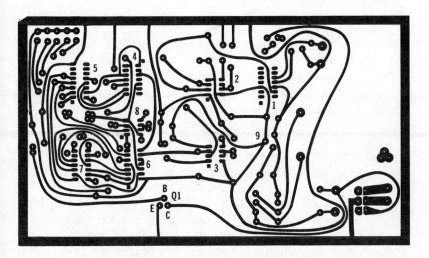

Fig. 13-4. Printed circuit layout shown half-size (foil side).

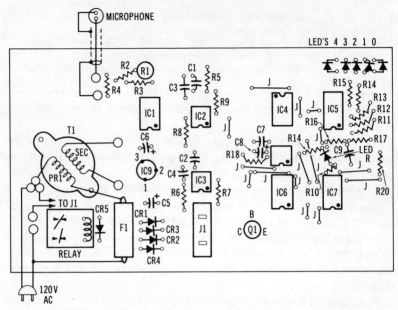

Fig. 13-5. Component layout shown from top of board. Fifteen jumper wires are required to complete the circuit.

TESTING AND ADJUSTING THE REMOTE CONTROL

The circuit can be tested and adjusted using only the built-in LED indicators and a dc voltmeter. The first check to be made is the timing of IC2. It will be helpful if you temporarily remove IC3 from the circuit to prevent it from resetting the counter while you perform the first part of the check.

Apply power to the unit and measure the voltage at pin 1 and pin 2 of IC2. The voltage at pin 1 should be about 12 volts and the voltage at pin 2 should be 5 ± 0.25 volt, measured with respect to ground. Set the sensitivity control about ¾ maximum clockwise. Gently tap the microphone while watching LEDR. It should light when the microphone is tapped, and remain lit for about 3 seconds. Each time the microphone is tapped LEDR should light for at least 2 seconds and not more than 4 seconds. It is important that the timing of IC2 falls into this range so that it will be able to sense each telephone ring separately. You may change the value of R8, if necessary, to bring the timing of IC2 within the range of 2 to 4 seconds.

To check the operation of IC4 and IC5, momentarily short pin 9 of IC6 to ground to clear the counter and set it to zero. LED0 should be lit.

Gently tap the microphone and wait for LEDR to be extinguished. When this occurs, LED1 should light, indicating that the counter has advanced one count. Connect a voltmeter between pin 3 of IC8 and ground. Tap the microphone while watching the voltmeter. At the end of the 3 seconds time period of IC2 the counter should advance to a count of 2, and the voltage at pin 3 of IC8 should rise from zero to about 5 volts. This voltage should hold for at least 25 and not more than 30 seconds. You may adjust R18 to bring the timing of IC8 into this range, if necessary.

After IC8 has completed its operation, tap the microphone two more times to advance the counter to 4. At this time IC8 should again be activated as indicated by the voltmeter connected to pin 3. When IC8 completes its second cycle, the relay should be activated.

You may check the operation of the reset circuitry by tapping the microphone three times to simulate three telephone rings without waiting the required 25-second delay. When this is done the counter should not advance to a count of 3, but should reset to zero after a count of 2.

Replace IC3 in its socket. Check the timing of IC3 by connecting a voltmeter to pin 3. The voltage should be zero before the circuit is triggered by tapping the microphone, and should rise to about 5 volts and hold for at least 80 and not more than 95 seconds. You may adjust R6, if necessary, to bring the timing into this range. The operation of IC3 may be visually checked by advancing the counter to a count of 1 and watching the LED indicators. After IC3 completes its 90-second time, the counter should be reset to zero.

The final adjustment is the sensitivity control. If possible have a friend call you up and let the telephone ring for about a minute. Locate the remote control microphone as close to the telephone bell as possible. Set the sensitivity control to the maximum counterclockwise (least sensitive) position. Slowly turn the sensitivity control while watching LEDR, and leave it set to the least sensitive position which gives a reliable detection of the sound. It is best to avoid excessive sensitivity so that the circuit does not respond to random noises in the house.

The remote control is now ready to be placed in operation by connecting the appliance to the receptacle on the circuit board and turning its power switch on. Once the hookup is made the appliance will automatically be turned on when you call your own telephone number with the proper code. The 25-second delay time between the 2nd and 3rd rings is not critical, but should not be less than 25 seconds and not more than 40 seconds. The circuit will operate properly if you happen to get a partial ring when you call, but if there is any doubt that the phone has rung it is best to hang up, wait 2 minutes, and try again. Also, since it is possible that you may call during a time that the circuit is in an activated state due to any calls made by others, it would be good practice to ring the code two times, spaced several minutes apart.

Chapter 14

Citizens Band Telephone Patch

For many years the radio amateur has taken advantage of the telephone to extend his voice communications beyond the physical location of his transmitter and receiver. Such an arrangement is called a telephone patch, and the Federal Communications Commission has seen fit to allow this procedure to be used with the Citizens band. As with all radio communications, strict rules specified by the FCC must be followed, and the telephone patch equipment interfaced with the telephone network must meet the requirements of Part 68 of the FCC Rules and Regulations. The telephone patch described here is a simple, easy to build circuit which will provide the means to connect and operate a CB radio with the telephone network. An outline of CB Rule 40 of the Citizens Band Service Rules appears at the end of this chapter.

The circuitry for the CB Telephone Patch (Fig. 14-1) is contained on a small printed circuit board measuring 2¾ inches × 4¾ inches (7 cm × 12 cm) and may be inserted into the cabinet of a CB radio (assuming there is sufficient room), to make this accessory a valuable addition to the radio. The circuit operates from a common 9-volt transistor battery, and should have an operating life equivalent to that provided by any of the small handheld radios. An additional feature of this telephone patch is a monitoring circuit which permits the CB operator to hear both sides of the telephone conversation.

ABOUT THE CIRCUIT

Refer to Fig. 14-2. The CB Telephone Patch consists of two sections of a quad operational amplifier chip, an audio power amplifier, and an

Fig. 14-1. CB Telephone Patch.

isolation transformer which connects the amplifier circuitry to the telephone line. The primary of T1 is connected to the telephone line through series capacitor C2. This provides dc blocking so that only audio frequencies are impressed upon the transformer winding. The circuit composed of R1, R2, C1, and Q1 is a holding circuit which provides a dc path for the telephone line current, which is necessary to hold the line in operation when the telephone set is on the hook. The ac impedance of this circuit is very high, due to the action of C1 which does not permit any audio frequencies to be impressed upon the base of Q1. This prevents the holding circuit from loading down the line, and preserves the impedance matching action of T1.

When the CB Telephone Patch is operated in transmit mode, S1 connects the transformer winding to operational amplifier IC1A through the transmit volume control, R5. IC1A has a gain of about 5, which provides sufficient drive to the microphone input of the CB radio through the Normal/Patch switch. Thus, any audio signals present on the telephone line will be fed to the microphone input circuit of the CB radio when the circuit is operated in transmit mode.

During receiving mode, S1 is thrown to the receive position. Audio information received by the CB radio and fed to its speaker is also impressed upon the input of operational amplifier IC1B through the receive volume control, R14. IC1B has a gain of about 5, and drives the transformer winding through R10. Thus, in receive mode, any audio signals received by the CB radio will be impressed upon the telephone line through T1.

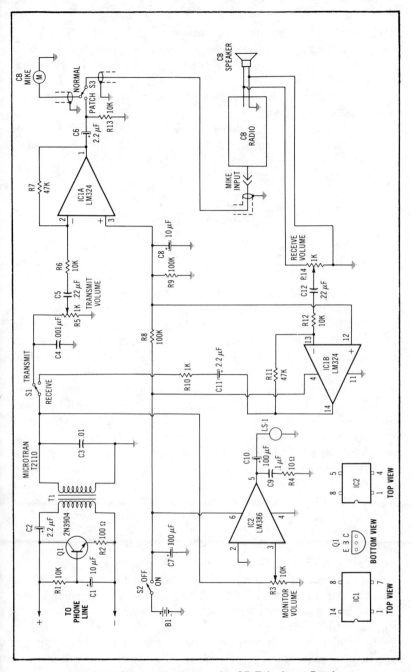

Fig. 14-2. Schematic diagram for CB Telephone Patch.

117

The use of a monitoring audio amplifier in the CB telephone patch permits hanging up the telephone once the call is initiated. The monitor amplifier will reproduce both sides of the conversation, and its volume can be set to a satisfactory level by means of the monitor volume control, R3. IC2 is a low-power audio amplifier which will drive a small loudspeaker.

Power for the circuit is provided by a common 9-volt transistor battery. Current drain on the battery will depend upon the level of volume provided by the monitor loudspeaker, and will be in the 20-milliampere range at low volume levels.

CONSTRUCTION

The parts list is given in Table 14-1. The full size layout for the single-sided printed circuit is shown in Fig. 14-3. Component placement, as seen from the component side of the board is shown in Fig. 14-4. Operating controls for the circuit, R3, R5, R14, S1, and S3 are shown mounted to the printed circuit board. This method of construction is left as an option of the builder. You may want to locate these controls on the front panel of a cabinet. In either case, the printed circuit layout as shown will be the same. Use 14 pin and 8 pin DIP sockets for IC1 and IC2. The small extra cost will be a good investment if the circuit should ever need service. Pay careful attention when mounting the electrolytic capacitors and Q1. These components will not work and may be damaged if they are inserted into the circuit board in the incorrect direction. This caution applies also to IC1 and IC2. Accidental reversal of the direction in which they are inserted into the board will almost certainly cause damage to the integrated circuit. Each IC has a small dot molded into the top of the plastic case next to pin one. Figs. 14-3 and 14-4 show clearly which pin of the printed circuit layout is pin 1.

The printed circuit layout shown in Fig. 14-3 provides proper mounting for T1 as specified in the parts list. It is not necessary that this exact component be used. If you substitute, use an audio transformer which has a 1 to 1 turns ratio and is rated at about 900-ohms impedance over the frequency range of 300 to 3000 Hz. You may have to alter the printed circuit layout for the transformer connections if you use a substitute transformer.

The 9-volt transistor battery which powers the circuit may be installed at any convenient location. If so desired, you may wish to make the printed circuit board slightly larger to accommodate the battery. If so,

Table 14-1. CB Telephone Patch Parts List

Item	Description
C1	10-μF, 25-volt tantalum capacitor
C2	2.2-μF, 25-volt tantalum capacitor
C3	.01-μF, disc capacitor
C4	1000-pF, disc capacitor
C5	.22-μF, 25-volt ceramic capacitor
C6	2.2-μF, 25-volt tantalum capacitor
C7	100-μF, 10-volt tantalum capacitor
C8	10-μF, 10-volt tantalum capacitor
C9	.1-μF, 25-volt ceramic capacitor
C10	100-μF, 10-volt tantalum capacitor
C11	2.2-μF, 25-volt tantalum capacitor
C12	.22-μF, 25-volt ceramic capacitor
R1	10K, ¼-watt, 10% composition resistor
R2	100-ohm, ¼-watt, 10% composition resistor
R3	10K potentiometer
R4	10-ohm, ¼-watt, 10% composition resistor
R5	1K potentiometer
R6	10K, ¼-watt, 10% composition resistor
R7	47K, ¼-watt, 10% composition resistor
R8	100K, ¼-watt, 10% composition resistor
R9	100K, ¼-watt, 10% composition resistor
R10	1K, ¼-watt, 10% composition resistor
R11	47K, ¼-watt, 10% composition resistor
R12	10K, ¼-watt, 10% composition resistor
R13	10K, ¼-watt, 10% composition resistor
R14	1K potentiometer
IC1	LM324 National Semiconductor
IC2	LM386 National Semiconductor
Q1	2N3904
S1, S3	Spdt toggle switch
S2	Spst toggle switch
T1	Microtran T2110 900-ohm-to-900 ohm telephone coupling transformer, or equivalent
LS1	4- or 8-ohm loudspeaker, 3 inch or larger
B1	9-volt transistor battery
P1	Telephone plug, 4 prong or modular

be sure to leave sufficient room to mount a clamp to hold the battery in place. Do not forget to install a power switch. This is shown in the schematic diagram as S2.

INSTALLATION

When connecting the CB Telephone Patch to your CB transceiver, follow the connections shown in the schematic diagram, Fig. 14-2.

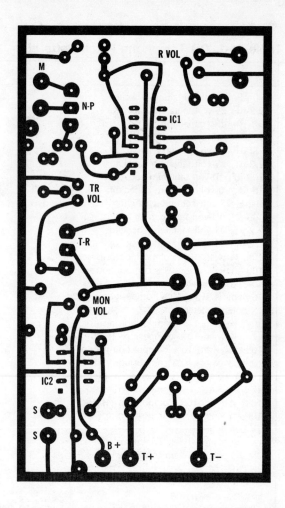

Fig. 14-3. Printed circuit layout shown full size (foil side).

There are two connections which must be made to the CB radio, not including ground wires. These are the microphone input jack and the speaker terminals. You may wish to purchase a microphone plug and jack which match those components on your CB radio. This will permit you to make the connections to the radio without disturbing either the microphone cable or the connections to the jack within the radio. Use shielded microphone wire for the microphone connections. Note that when S3 is set to the normal position, the microphone will be connected to the CB radio in the usual manner, permitting normal operation. When making the connection to the speaker of the CB radio, you may

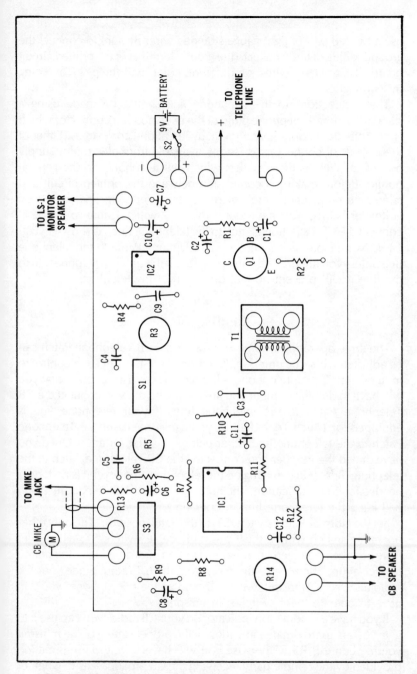

Fig. 14-4. Component layout shown from component side of board.

use a twisted pair of light gauge stranded wire. Be sure to connect the grounded side of the speaker to the ground terminal of the printed circuit board. Using two wires of different color will help in avoiding misconnections.

The connections to the telephone line should be made using a standard 4-prong telephone plug, or the new modular connector which is now standard equipment with many telephone companies. Either of these types of connectors is readily available from electronics supply houses, as well as those outlets which sell telephones to the general public. Before making a connection between the printed circuit and telephone plug, check the polarity of the telephone line with a dc voltmeter. Once you have done this you will be able to properly connect the CB Telephone Patch to the telephone line, as shown in Fig. 14-4. *Note:* Do not connect the CB Telephone Patch to the telephone line unless you are placing it in operation. If the unit is left connected to the line it will prevent normal operation of the telephone.

OPERATION

The three operating volume controls in the CB Telephone Patch can be adjusted while the unit is in operation. As a start, set each control to midposition. If your telephone is the type which uses a rotary dial you will have to dial the number you are calling before you connect the CB Telephone Patch to the telephone line. If your telephone line is equipped for Touch Tone dialing, you can connect the CB Telephone Patch to the telephone line either before you dial, or after. Once you have dialed the number and connected the CB Telephone Patch to the telephone line, you can hang up the telephone whenever it is no longer needed for communication. The holding circuit of CB Telephone Patch will keep the telephone line in operation.

Set the Normal/Patch switch, S3, to the Patch position. The only other operating switch will be the Transmit/Receive switch, S1, which will have to be operated manually as you transmit the telephone conversation, and receive the reply from the CB radio. Since your CB radio also has a Transmit/Receive button on the microphone, you will have to operate these switches in unison.

If you have a modulation indicator on your CB radio, you can use it to give you an approximate indication of the proper setting of the transmit volume control, R5. Otherwise you will have to rely on information provided by the remote party receiving the CB broadcast as to whether or not the level of modulation is sufficient.

122

Before adjusting the receive volume control, R14, set your CB radio volume control for a comfortable volume setting. Use the squelch control to eliminate receiver noise when no signal is present. By speaking into your telephone handset, you can ask the party at the other end of the telephone line if the received volume from the CB radio is sufficient. This will enable you to adjust R14 for proper volume on the telephone line signal.

Once the receive and transmit volume controls are set to a satisfactory volume level for both parties, you can hang up your telephone and use the monitor feature of CB Telephone Patch to hear both sides of the conversation as you operate the receive/transmit switches. Set the monitor volume control, R3, for a comfortable volume setting.

After you have used your CB Telephone Patch for the first time, you can leave the three volume control settings as they are. The next time you operate the unit they will probably need little or no readjustment. When the telephone patch conversation is terminated, be sure to disconnect the CB Telephone Patch from the telephone line. Failure to do this will result in an inoperative telephone.

Before placing the CB Telephone Patch on the air, read and obey the following rules which are spelled out under CB Rule 40 of the Citizens Band Radio Service, effective August 1, 1978:

(a) You may connect your CB transmitter to a telephone if you comply with *all* of the following:

   (1) You, or someone authorized to operate under your license, must be present at your CB station and must—

      (i) Manually make the connection (the connection must not be made by remote control);
      (ii) Supervise the operation of the transmitter during the connection;
      (iii) Listen to each communication during the connection; and
      (iv) Stop all communications if there are operations in violation of these rules.

   (2) Each communication during the telephone connection must comply with all of these rules.

   (3) You must obey any restriction that the telephone company places on the connection of a CB transmitter to a telephone.

(b) The CB transmitter you connect to a telephone must not be shared with any other CB station.

(c) If you connect your CB transmitter to a telephone, you must use a phone patch device which has been registered with the FCC.

Chapter 15

Conferencer

Today it is common for many people to have more than one telephone line coming into the house. Usually the second line is called a "teen-age telephone" and is provided by the telephone company at a special reduced rate. While the second telephone allows young people to talk on the telephone for hours at a time without tying up the first telephone, there is an additional benefit which can be derived by taking advantage of the fact that two independent telephone lines are available. This is the conference call which has been available to commercial businesses for years. Of course, the telephone company would be glad to equip your telephone for conference call capability, but they will want to charge you a monthly tariff whether you use it or not. If you build the Conferencer (Fig. 15-1), you can add conference call capability to your telephone and you won't have to pay for its use. Ma Bell will not even know that you have it, since it has no effect on telephone performance, except to allow simultaneous communications between your two telephone lines and two others. The Conferencer can be built at very low cost, making it a valuable addition to your telephone. It requires no external power source for its operation and is always ready for use whenever you are. The only requirement for making use of the Conferencer is to have two independent telephone lines available at a single location in your home or business.

ABOUT THE CIRCUIT

In order to connect two independent telephone lines together without disturbing the dc conditions on either line, a transformer must be used.

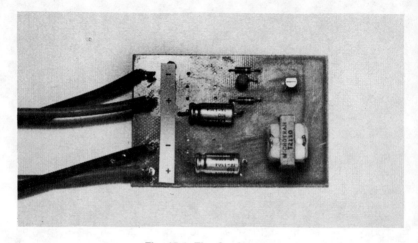

Fig. 15-1. The Conferencer.

This is the heart of the circuit which allows audio information on one line to be impressed on another line with no dc connection between them. T1 is a telephone line coupling transformer which has been designed for telephone line use. As shown in the schematic diagram of the Conferencer (Fig. 15-2), each telephone line is connected to separate windings of T1 through coupling capacitors.

Telephone line No. 1 is connected across one winding of the transformer through S1 and electrolytic capacitor C1. The capacitor prevents any dc from flowing through the transformer winding, while coupling the audio frequency information appearing across the telephone line to the transformer winding. The dc isolation provided by C1 ensures that the normal dc conditions of the telephone line are not disturbed. In a similar manner telephone line number two is connected to the second winding of the transformer. Each telephone line can be independently controlled through the action of S1 and S2, which have been provided to allow connection and disconnection of each telephone line as required.

An additional circuit has been included in the second telephone line circuit. This is a dc holding circuit which provides the proper dc characteristic to hold telephone line number two in operation even though no telephone on that line is present. The circuit composed of R1, R2, C3, and Q1 acts like a resistance to dc, and as a high impedance to audio signals. The high impedance of the circuit is provided by C3,

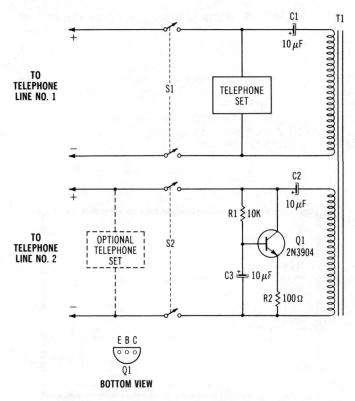

Fig. 15-2. Schematic diagram of the Conferencer.

which prevents any audio signals from appearing at the base of Q1. Thus, any audio voltage appearing across telephone line No. 2 will not cause a corresponding current in Q1.

Note that only one telephone is required when using The Conferencer. If that telephone is equipped for Touch Tone dialing, it will be possible to dial telephone numbers to either line as required. If only rotary dialing is available and it is desirable to dial a number on line No. 2, then a second telephone can be connected across that line for dialing purposes only. This is shown on the schematic diagram as an optional telephone set.

CONSTRUCTION

The parts list is given in Table 15-1. The printed circuit for the Conferencer as shown full size in Fig. 15-3 measures 2 inches × 3

Table 15-1. Parts List for the Conferencer

Item	Description
C1	10-μF, 50-volt electrolytic capacitor
C2	10-μF, 50-volt electrolytic capacitor
C3	10-μF, 15-volt tantalum capacitor
R1	10K, ¼-watt, 10% composition resistor
R2	100-ohm, ¼-watt, 10% composition resistor
Q1	2N3904 npn silicon transistor
S1	Dpst toggle switch
S2	Dpst toggle switch
T1	Telephone coupling transformer, Microtran T2110 or equivalent

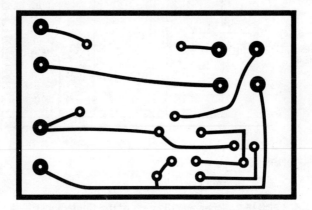

Fig. 15-3. Printed circuit layout shown full size (foil side).

inches (5 cm × 7.6 cm). This size is based on using the transformer specified in the parts list. For this application you can use any audio coupling transformer which is rated at about 900 ohms both primary and secondary, over the frequency range of about 300 to 3000 Hz. Be sure to allow extra room on the printed circuit board for the substitute transformer, if necessary.

When mounting components to the printed circuit board follow the layout shown in Fig. 15-4 which is a view of the board shown from the component side. Pay careful attention when inserting Q1 into the board so that the three leads are placed in the proper holes. The same caution applies to C1 and C2 which are polarized components. If these capacitors are inadvertently placed in the board incorrectly, the telephone line dc conditions will be disturbed.

There is no provision to mount S1 and S2 on the printed circuit board.

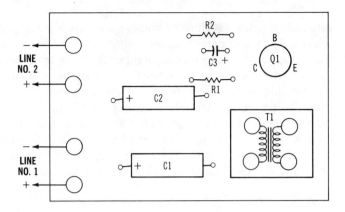

Fig. 15-4. Component layout shown from component side of board.

The placement of these switches is left as an option of the builder. You may want to place the entire circuit within the confines of a telephone set if there is room, or you may want to house The Conferencer in a small cabinet to be located next to the telephone. When making connections to the telephone line, be sure to follow correct procedure and use the standard four prong telephone plugs or the new modular connectors which are available from many electronics parts suppliers. These connectors allow easy connection and disconnection to the telephone line, which is an important factor should the unit ever require servicing.

When making connections between the printed circuit and the telephone lines it will be necessary to observe the polarity of the telephone line. Use a dc voltmeter to determine the polarity of each line before wiring the telephone plugs. You must maintain correct polarity when connecting a Touch Tone telephone to the line and it is suggested to follow the original wiring of the telephone set after it is disconnected to install the switch, S1. Bear in mind that an optional telephone set connected across line No. 2 is necessary only if you do not have Touch Tone dialing, and you wish to dial numbers on line number two.

OPERATION OF THE CONFERENCER

You will be able to make and receive calls on either telephone line by closing the appropriate switch and using the telephone set connected across line No. 1. If you wish to initiate a conference call, use the following procedure:

1. Close S1 and make a call on line No. 1. When you have

established contact with party number one, ask him/her to hold while you contact party number two.

2. Close S2, and dial the second number from the telephone set connected across line No. 1. When the second party answers, you will be able to carry on a three way conversation.

3. When using only rotary dial telephones, perform step two by dialing the second number from the optional telephone set connected across line No. 2 before S2 is closed. Once the number is dialed, you can close S2 and hang up telephone set number two and carry on the conference call in the normal manner.

Chapter 16

Telephone Beacon

Have you ever had a situation where, for one reason or another, the sound of the telephone bell was either masked by a high level of ambient noise, or was too faint to be heard due to a great distance between you and the bell? The Telephone Beacon (Fig. 16-1) has been designed for such a situation. With this telephone accessory you will be able to use an ordinary 120-volt incandescent lamp, of any wattage, as a visual signal with which all telephone calls are announced. This is a simple unit which can be constructed at very low cost, and can be

Fig. 16-1. The Telephone Beacon.

placed anywhere inside or outside your home or business wherever you care to run a pair of wires. It is powered by the 120-volt ac power line, and has infinite isolation between the power line and telephone circuit. It operates only during the time that the ringing signal appears across your telephone line, and at all other times it presents an infinite impedance to the telephone network. Thus, it cannot be detected by the telephone company, nor will it affect telephone performance. In addition to its obvious benefits to those who cannot hear the ordinary telephone bell because of ambient noise levels, the Telephone Beacon can be a boon to people who are hard of hearing. The Telephone Beacon is completely solid state in its design, making use of an alternating current control device, called a triac, to perform the ac power switching function.

ABOUT THE CIRCUIT

The heart of the Telephone Beacon is a triac, which is a generic term that has been coined to identify a three electrode semiconductor switch which is triggered into conduction in response to a gate signal. The action of the triac is similar to that of a silicon controlled rectifier (SCR), except that it is able to conduct current in both directions as required in an ac circuit. As shown in the schematic diagram (Fig. 16-2) the main conducting terminals of the triac, T1 and T2, are connected in series with an ordinary incandescent lamp and the ac power line. When a signal is impressed upon the third lead, the gate, the triac is triggered

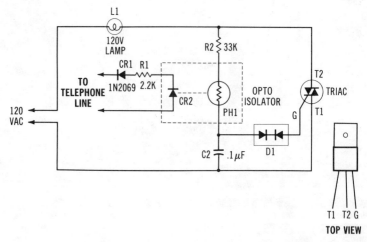

Fig. 16-2. Schematic diagram of the Telephone Beacon.

into conduction during that portion of the alternating current sine wave. When the current through the triac passes through zero, the device shuts off unless the gate signal remains impressed upon the gate terminal. Thus the triac can be made to operate as a solid-state ac power switch by the action of a relatively low current gate signal applied to the gate electrode.

In order to operate an ac powered device by signals coming from the telephone line, some means of connection must be used so that there is complete isolation between the telephone network and power line. This is accomplished by another solid-state device called an optoisolator. The optoisolator used in the Telephone Beacon consists of an ordinary LED (light-emitting diode) coupled to a photocell. Such a device is available as a complete assembly, or can be easily constructed by placing an LED and a photocell in a small enclosure.

The photocell of the optoisolator, along with a resistor, capacitor, and diac trigger element is connected to the gate circuit of the triac. When no light is impressed upon the photocell, its resistance is extremely high, resulting in virtually no voltage being impressed upon the diac. This element acts somewhat like a zener diode which can conduct current in both directions when its breakdown voltage is reached. With insufficient voltage impressed upon the diac, it acts like an open circuit and prevents any current flow in or out of the gate terminal of the triac. Thus no current flows through the lamp.

The LED component of the optoisolator is connected in series with a silicon diode and resistor. When the telephone is idle, the normal 48-volt dc potential appearing across the telephone line is impressed across CR1 with a polarity which opposes the forward current direction of the diode. No current flows, and the LED is thus extinguished. If a ringing signal should appear across the telephone line, the 200 volt peak-to-peak ringing signal has sufficient amplitude to overcome the dc voltage which cuts off CR1. This causes current to flow through R1 and CR2 so that some light is generated by the LED. Since CR2 and the photocell are placed in close proximity, the light falling upon the photocell is sufficient to cause its resistance to decrease to a very low value. The lowered value of resistance provides sufficient voltage to cause the diac to conduct, turning on the triac. The lamp connected in series with the triac is now lighted in response to the telephone signal appearing across the telephone line, and will turn off and on with the same repetition rate as the telephone bell. When the ringing signal no longer appears across the telephone line the circuit resumes its standby state, cutting off all power to the lamp.

CONSTRUCTION

The parts list is given in Table 16-1. The entire circuit for the Telephone Beacon, with the exception of the incandescent lamp, is contained on a small printed circuit board measuring only 2 inches × 2½ inches (5 cm × 6.35 cm). The full-scale layout of the printed circuit board as shown from the copper side of the board is illustrated in Fig. 16-3, and the parts layout as seen from the component side of the board is shown in Fig. 16-4. The triac specified in the parts list is in a 6-ampere TO-220 (plastic) case style. This part may be substituted with any triac capable of carrying the current of the lamp you plan to use, and if the substitute part is the stud mounted type you can easily alter the printed circuit layout as necessary. When laying out the copper paths, be sure to adhere to the width of the heavy conductors as shown in Fig. 16-3. These paths will carry the relatively heavy current demanded by the

Table 16-1. Telephone Beacon Parts List

Item	Description
R1	2.2K, ¼-watt, 10% composition resistor
R2	33K, ¼-watt, 10% composition resistor
C1	0.1-μF, 500-volt disc capacitor
CR1	1N2069 silicon diode
CR2	LED (Part of optoisolator Vactec VTL5C1)
D1	1N5758 diac
PH1	Photocell (Part of optoisolator VTL5C1)
T1	Triac, 200-volt, 6-ampere RCA T2500B or equivalent
L1	Incandescent lamp, 120 volt, 100 watts or less

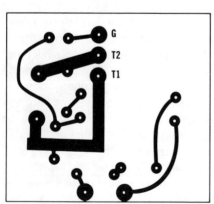

Fig. 16-3. Printed circuit layout shown full size (foil side).

incandescent lamp. For lamps of 100 watts or less you can mount the triac to the printed circuit board with no heat sink, since the triac will be operating at a current level much lower than its maximum rating. The intermittent nature of the telephone ringing signal (2 seconds on, 4 seconds off) also contributes to a cool operating triac. Since this circuit operates directly from the ac power line with no isolation transformer, be very careful that no part of the circuit comes in contact with any housing in which you may place the circuit board.

You may purchase a ready made optoisolator as specified in the parts list, or you can construct one by placing a photocell and an LED in a small light proof tube and sealing the ends so that no light can enter. Set the LED as close to the sensitive area of the photocell as possible for best performance. Fig. 16-5 illustrates the construction details of such an optoisolator.

When mounting CR1, CR2, and the triac to the board be sure to follow the direction for these polarized components as shown in Fig. 16-4. The triac is mounted to the board with the metal mounting flange placed against the board. Do not let the mounting screw touch anything, since it is electrically in contact with one of the leads of the triac. The diac is a nonpolarized device, and can be placed in the printed circuit board in any direction. The same holds true for the photocell.

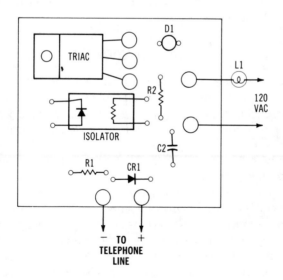

Fig. 16-4. Parts layout shown from component side of board.

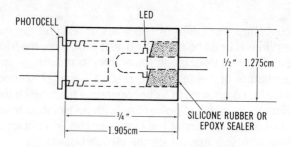

Fig. 16-5. Construction details of optoisolator.

After you have completed the wiring of the unit you can check it out before connecting it to the telephone line to be sure it operates properly. To make this check you will need a dc supply capable of driving about 10 milliamperes through CR2. This will take about 24 volts dc. Connect the dc supply to the telephone line terminals of the printed circuit board in the *opposite* direction as shown in Fig. 16-4 and the schematic diagram. The purpose of connecting the dc supply to the unit in this manner is to cause current flow through the LED. Connect a 120-volt lamp in series with the ac terminals of the printed circuit board as illustrated in Fig. 16-4, and apply 120 volts ac to the circuit. The lamp should light. Disconnect the dc supply from the telephone line terminals of the printed circuit board. The lamp should extinguish. Once this test is completed, you will be ready to connect the Telephone Beacon to the telephone line.

INSTALLATION

Connections to the telephone line should be made using a standard 4-prong telephone plug, or modular plug. Before wiring up the connector, check the polarity of your telephone line with a dc voltmeter capable of measuring 50 volts or more. Observe correct polarity when wiring the connector to the printed circuit, as shown in Fig. 16-4. If you happen to wire the circuit with the incorrect polarity, you will have an inoperative telephone line.

Use good wiring practice when wiring the lamp, power line cord, and printed circuit board as shown in the schematic. Due to the nature of the circuit, the lamp will always limit the current drawn from the power line to a safe value in the event that any of the components of the printed circuit board should become defective. In addition, the telephone line will never be subject to hazardous voltages, since it is protected by the isolation property of the optoisolator.

Chapter 17

Automatic Bell Silencer

The telephone is a wonderful device which touches us all, and life as we know it couldn't exist without it. However, sometimes it can cause annoyance in our lives when the attention-getting bell rings at the wrong time. Have you ever just begun to enjoy an evening meal, only to be interrupted by that call which comes at the wrong time? The Automatic Bell Silencer (Fig. 17-1) described here is the answer to such a problem. By the touch of a button, you can place your telephone in limbo for a specified length of time. At the end of this time period the telephone is automatically reverted to normal operation. Anyone who attempts to call you during the silent period will receive a busy signal, so that in most cases the call will be attempted a few minutes later when it will be completed. A reset button has been provided which allows cancellation of the timed cycle at any time. The Automatic Bell Silencer has a distinct advantage over taking the telephone off the hook when it is desired not to have any calls. It is too easy to forget to replace the telephone handset. This may result in the loss of several calls.

The circuit is powered by a common 9-volt transistor battery which should last well over a year, and a built-in LED provides positive indication that the silencer is operating properly. The circuit operates by "fooling" the telephone line into thinking that your telephone is in operation. This prevents the bell from operating, and signals the busy signal to others who call you during the silent period. Since the Automatic Bell Silencer places a resistive load, equivalent to your telephone set, across the telephone line, it will not harm the telephone network nor cause improper operation of your telephone. The

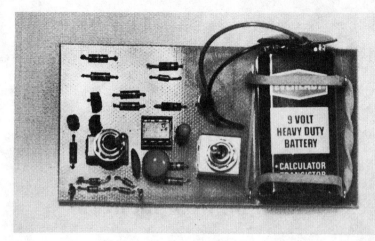

Fig. 17-1. Automatic Bell Silencer.

Automatic Bell Silencer is not recommended for two or four party telephone lines, since it will prevent telephone calls from reaching the other parties as well as your own.

ABOUT THE CIRCUIT

The heart of the Automatic Bell Silencer is a 555 timer integrated circuit hooked up into a "suicide" circuit. This means that once the timing cycle of the integrated circuit is initiated, the circuit is doomed to kill itself when the time interval is completed.

Refer to Fig. 17-2. IC1 is connected as a one-shot or monostable multivibrator. The circuit is energized by means of a double-pole switch, S1. Pressing this spring-loaded switch causes battery voltage to be applied to the power input pin of IC1, pin 8, and grounds the trigger terminal, pin 2. This action initiates the timed cycle of IC1 which is determined by the RC time constant composed of R3 and C3. For the values shown on the schematic, this time interval is about 12 minutes.

The output terminal of IC1, pin 3, goes to almost full battery voltage during the timed cycle. This voltage is fed to the base of transistors Q2 and Q3 through current-limiting resistors. Both transistors become saturated, resulting in essentially zero voltage at their collectors. The voltage at the collector of Q2 is fed to the base of Q1 through R6. This causes Q1 to become saturated. Since the emitter-collector of Q1 is connected across S1A, power of IC1 is maintained even though S1 is released and returns to the off position.

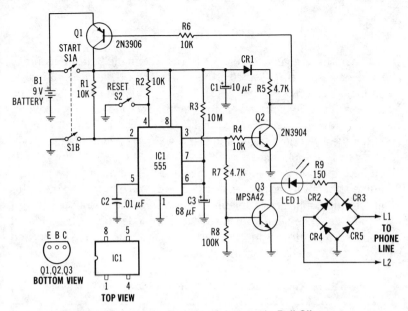

Fig. 17-2. Schematic diagram of Automatic Bell Silencer.

The collector of Q3 is connected to the telephone line thrugh LED1, R9, and a diode bridge. This essentially places R9 across the telephone line, and simulates a real telephone being off the hook. The telephone line is now busy, and the bell signal is prevented from being impressed across the line by the telephone equipment at the central office. The LED indicator is illuminated by the dc being drawn from the telephone line. This provides positive indication that the Automatic Bell Silencer is in operation.

Looking back at timer IC1, capacitor C3 begins charging the moment the timed cycle is initiated. When the voltage across C3 reaches ⅔ the battery voltage, the timing cycle of IC1 is completed and pin 3 goes to zero volts. This cuts off Q2, Q3, and also Q1. Thus, the entire circuit is deprived of battery power, and is dead. When Q3 cuts off, the resistive load of R9 is removed from the line, placing the telephone back in normal operation. Diode CR1 has been placed in the circuit to prevent a sneak path of current from the battery through IC1 when the circuit is not in operation. This results in virtually zero current being supplied by the battery during the off time, and provides long battery life.

In the event that the timed cycle needs to be cancelled before it is automatically completed, a reset switch has been provided in the circuit. This switch grounds out the reset terminal of IC1, pin 4, and causes the timer to revert back to the off condition.

CONSTRUCTION

The parts list is given in Table 17-1. The Automatic Bell Silencer may be built on a small single-sided printed circuit board measuring 2 inches × 4 inches (5 cm × 10 cm). This includes room to mount the 9-volt transistor battery which powers the circuit. The printed circuit layout, shown in full size, is illustrated in Fig. 17-3. The reverse side of the board is the component side, and is shown from that view in Fig. 17-4. As shown in the photograph of the unit (Fig. 17-1), all components, including the two operating switches and LED indicator, have been mounted directly to the printed circuit board. You may wish to build this unit into the confines of a telephone, and have the operating controls and indicator accessible from the outside. This is left up to the builder. It is a simple matter to locate these parts away from the printed circuit and connect them with light gauge wires.

Table 17-1. Automatic Bell Silencer Parts List

Item	Description
C1	10-μF, 25-volt electrolytic capacitor
C2	0.01-μF, disc capacitor
C3	68-μF, 10-volt tantalum capacitor
CR1	1N4148 silicon diode
CR2	1N4148 silicon diode
CR3	1N4148 silicon diode
CR4	1N4148 silicon diode
CR5	1N4148 silicon diode
R1	10K, ¼-watt, 10% composition resistor
R2	10K, ¼-watt, 10% composition resistor
R3	10-megohm, ¼-watt, 10% composition resistor
R4	10K, ¼-watt, 10% composition resistor
R5	4.7K, ¼-watt, 10% composition resistor
R6	10K, ¼-watt, 10% composition resistor
R7	4.7K, ¼-watt, 10% composition resistor
R8	100K, ¼-watt, 10% composition resistor
R9	150-ohm, ¼-watt, 10% compostion resistor
LED1	Light-emitting diode
IC1	555 timer
Q1	2N3906 or equivalent
Q2	2N3904 or equivalent
Q3	MPSA42 Motorola or equivalent
Battery	9-volt NEDA 1604D
Battery Connector	H.H. Smith No. 1234
S1	Dpst spring-return toggle switch, C & K 7208 or equivalent
S2	Spst spring-return toggle switch, C & K 7108 or equivalent

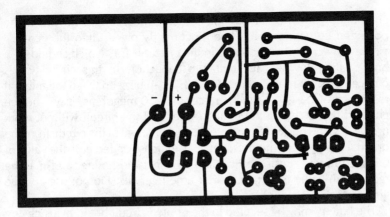

Fig. 17-3. Printed circuit layout shown full size (foil side).

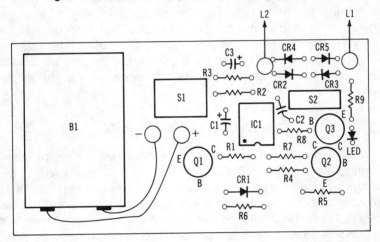

Fig. 17-4. Component layout shown from component side of board.

Although this is a simple circuit which should require no service once it is built, it is recommended that an IC socket be used for IC1. The cost is minimal, and it makes a more professional job.

Before mounting the electrolytic capacitors, diodes, and LED, consult Fig. 17-4 for the proper orientation of these parts. This cannot be overstressed, since one incorrectly placed part will render the circuit inoperative. The same advice applies when mounting the transistors. Be careful to place the emitter, base, and collector leads in the proper holes as shown in Fig. 17-4. Bear in mind that this circuit uses three different types of transistors. Place them in the correct locations. Should the

141

circuit not work once it is assembled, it is almost certain that one of the polarized components has been incorrectly placed into the board.

The only critical component in the circuit is C3, which should be a high-quality, low-leakage tantalum capacitor. In this circuit C3 is charged by a 10-megohm resistor, R3. It can be seen that if the internal resistance of C3 is not much greater than 10 megohms, it will not be charged by R3 during the timing cycle, and the circuit will not work properly. If it is desired to change the time interval of the circuit from the original design of 12 minutes, you may increase or decrease the value of C3 accordingly. Resistor R3 will also affect the timing, and may be reduced in value to obtain shorter time intervals. Do not use a value greater than 10 megohms for R3.

The battery may be secured to the printed circuit board by using any method your ingenuity can dream up. The battery shown in the photograph has been secured by soldering four loops of wires to the board, with a small rubber band held captive. The battery can then be slipped under the rubber band where it will stay in place. Use a readily available battery connector to make connection between the printed circuit and battery. Be sure to wire the connector with the proper polarity.

CHECKOUT AND OPERATION

You can precheck the operation of the circuit using a dc voltmeter. Connect the negative lead of the voltmeter to the negative lead of the battery, and hold the positive lead of the voltmeter against pin 3 of IC1. Be careful not to short out adjacent pins of the IC. Actuate the Start switch. The meter should indicate that the voltage at pin 3 went from zero to about 8 volts when the start button was pressed. Throw the Reset button while monitoring the voltage at pin 3 of IC1. This voltage should immediately go to zero. Once this test has been made, the Automatic Bell Silencer is ready to be connected to the telephone line. The terminals marked L1 and L2 as shown in Fig. 17-4 are connected to L1 and L2 of the telephone line. These terminals are readily available at the encapsulated assembly within the telephone housing. There is no need to be concerned about polarity when making the connection, since the diode bridge in the unit will automatically feed the correct polarity to the circuit.

Once the Automatic Bell Silencer is connected to the telephone line you can press the start switch. The LED should light, indicating that the bell is now inoperative. If desired, you may wish to leave the circuit in operation to measure the timing interval. The LED will become extinguished at the end of this time. Otherwise, press the reset button to allow your telephone to receive any incoming telephone calls.

Chapter 18

The Babysitter

Usually you think of your telephone as a useful device only when you are home and able to make and receive telephone calls. However, it is possible to take advantage of the telephone even when you are not at home, no matter how far away you are. The Babysitter (Fig. 18-1) will equip your telephone for remote listening, so that you can check up on what is happening in your house any time you are away. Of course, The Babysitter cannot take the place of a real person, but it will give you the ability to call your own telephone and listen to the sounds received by a microphone placed anywhere you desire. The Babysitter will answer your telephone, couple a microphone amplifier to the telephone line for a specified length of time selected by you, and "hang up" at the end of that time ready to receive another telephone call. All this is done without any modifications to your telephone. The Babysitter is completely solid state, uses no relays, and is powered by the ac power line.

ABOUT THE CIRCUIT

Refer to Fig. 18-2. In order to answer a telephone call, the circuit has been designed to respond to the 90-volt, 20-Hz ringing signal. This section of The Babysitter is composed of IC1 and its associated components. Two NAND gate sections of IC1, sections A and B, are connected in a configuration called a latch circuit or flip-flop. This circuit has two stable states, and the output terminals, pins 3 and 4 of IC1, always assume opposite logic levels. When The Babysitter is in

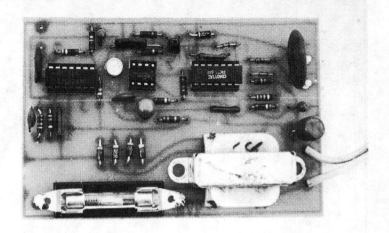

Fig. 18-1. The Babysitter.

standby mode, pin 3 of IC1 is at a logic zero level, and pin 4 is at a logic one level. These levels are represented by voltages of about zero and eight volts respectively, and the circuit will remain in this state until a negative-going pulse is impressed on terminal 1 of IC1A. This pulse will be the 20-Hz ringing signal which appears across the telephone line when an incoming call is received. Isolation between the telephone line and the ring detector is provided by C1 and R1. Since the ringing signal has an amplitude of about 200 volts peak-to-peak, a high value of isolation impedance is permissible. This also prevents any disturbance to the telephone line, and prevents the telephone company from detecting the additional circuit connected to the line. Diodes CR1 and CR2 prevent the voltage impressed upon terminal 1 of IC1A from exceeding the range of zero to eight volts.

The remaining two gates of IC1 are connected in a logic configuration which has been designed to ensure that the logic state of IC1A and IC1B always assumes the correct mode when the circuit is turned on. A relatively slow charging circuit composed of R16 and C11 holds pin 12 of IC1D to a zero logic level for about 1 second when power is applied to the circuit. This guarantees that pin 6 of IC1B is held to logic zero level during that time, and sets the latch circuit to its standby mode.

The ring-detector circuit responds to the ringing signal by causing the latch circuit of IC1 to assume its opposite stable state. Thus, the logic level at pin 4 changes from one to zero when a call is received. The

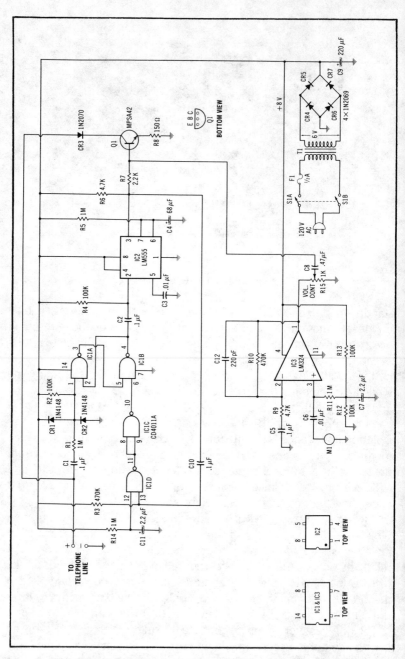

Fig. 18-2. Schematic diagram of The Babysitter.

146

negative-going transition at pin 4 is impressed upon the trigger input of IC2, which is a timer IC connected as a one-shot or monostable multivibrator. When IC2 is triggered, its output terminal, pin 3, changes from a voltage of zero to about eight volts. IC2 will remain in this state for a specified length of time determined by the RC time constant composed of R5 and C4. For the values shown in the schematic diagram, this time interval is about 70 seconds.

The output voltage of IC2 is impressed upon the base of Q1 through R8, and causes this transistor to conduct. This provides a dc path for the telephone line voltage, and essentially connects R8 across the telephone line. This, in effect, is "answering the telephone." At the end of the timed cycle of IC2, Q1 becomes cut off. This, in effect is "hanging up."

IC3 is one section of an operational amplifier integrated circuit which is connected as a high gain microphone amplifier with a voltage amplification of about 100. The output of this amplifier is impressed upon the base of Q1 through a volume control potentiometer so that any sounds reaching the microphone are impressed upon the telephone line through Q1. Thus, while IC2 holds Q1 in saturation during its timed cycle, the microphone amplifier output appears across the telephone line and can be heard by the calling party.

When IC2 reaches the end of its cycle, output terminal pin 3 goes to zero volts. This negative transition is impressed upon pin 6 of IC1A and causes the latch circuit to revert back to its original state. The circuit is now ready to receive the next telephone call.

Power for the circuit is provided by a common 6-volt filament transformer feeding a full-wave rectifier bridge. The output voltage is filtered by C9, and is about 8 volts dc. A fuse has been placed in the primary circuit of the transformer. This is good practice with any power-line operated device.

CONSTRUCTION

The parts list is given in Table 18-1. The Babysitter is constructed on a printed circuit board measuring about 4½ inches × 3 inches (11 cm × 7 cm), including the line operated power transformer and fuse block. A full-size view of the printed circuit, as shown from the copper side of the board is illustrated in Fig. 18-3. The component layout shown from the component side of the board is illustrated in Fig. 18-4. These two illustrations, along with the schematic diagram will provide the builder with all the necessary information needed to construct the unit.

Table 18-1. The Babysitter Parts List

Item	Description
C1	0.1-μF, 500-volt ceramic disc capacitor
C2	0.1-μF, 25-volt ceramic capacitor
C3	0.01-μF, 25-volt ceramic disc capacitor
C4	68-μF, 10-volt tantalum capacitor
C5	0.1-μF, 25-volt ceramic capacitor
C6	0.01-μF, 25-volt ceramic disc capacitor
C7	2.2-μF, 10-volt tantalum capacitor
C8	0.47-μF, 25-volt ceramic capacitor
C9	220-μF, 25-volt tantalum capacitor
C10	0.1-μF, 25-volt ceramic capacitor
C11	2.2-μF, 10-volt tantalum capacitor
C12	220-pF, 25-volt ceramic disc capacitor
CR1	1N4148 silicon diode or equivalent
CR2	1N4148 silicon diode or equivalent
CR3	1N2070 silicon diode or equivalent
CR4	1N2069 silicon diode or equivalent
CR5	1N2069 silicon diode or equivalent
CR6	1N2069 silicon diode or equivalent
CR7	1N2069 silicon diode or equivalent
IC1	RCA CD4011AE
IC2	National LM555
IC3	National LM324
Q1	MPSA42 Silicon npn Transistor or equivalent
R1	1-megohm, ¼-watt, 10% composition resistor
R2	100K, ¼-watt, 10% composition resistor
R3	470K, ¼-watt, 10% composition resistor
R4	100K, ¼-watt, 10% composition resistor
R5	1-megohm, ¼-watt, 10% composition resistor
R6	4.7K, ¼ watt, 10% composition resistor
R7	2.2K, ¼-watt, 10% composition resistor
R8	150-ohm, ¼-watt, 10% composition resistor
R9	4.7K ¼-watt, 10% composition resistor
R10	470K, ¼-watt, 10% composition resistor
R11	1-megohm, ¼-watt, 10% composition resistor
R12	100K, ¼-watt, 10% composition resistor
R13	100K, ¼-watt, 10% composition resistor
R14	1-megohm, ¼-watt, 10% composition resistor
R15	1K potentiometer, printed circuit mount
T1	6-volt transformer, Radio Shack 273-1384 or equivalent
F1	½ ampere slow-blow fuse
M1	Crystal or ceramic microphone
S1	Dpst toggle switch

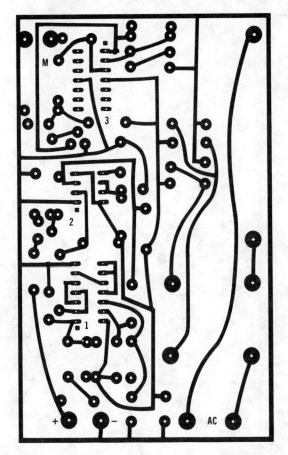

Fig. 18-3. Printed circuit layout shown full size (foil side).

The printed circuit layout has been designed to accommodate the miniature 6-volt power transformer which is specified in the parts list. This component is not at all critical, and if you plan to use a substitute 6-volt transformer be sure to allow sufficient room on your printed circuit board and alter the printed wiring accordingly. Be sure to include a fuse in the primary circuit of the transformer as shown in the schematic diagram.

Although this unit operates at a low voltage of about 8 volts dc, there are three components which must be able to withstand the relatively high voltage pulses which appear across the telephone line during the ringing signal. These parts are C1, CR3, and Q1. Use the parts specified

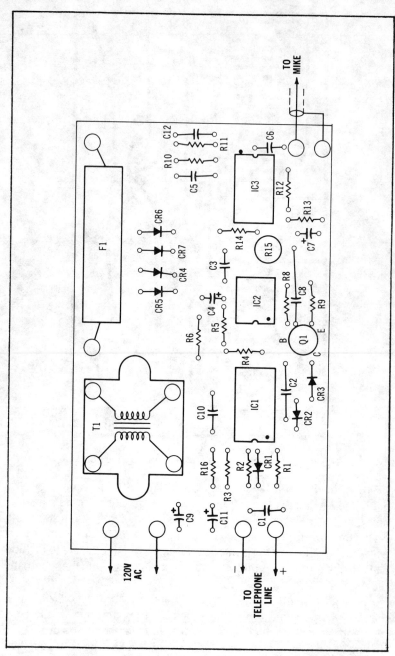

Fig. 18-4. Component layout shown from component side of board.

150

in the parts list if possible. If substitution is necessary be sure that the substituted components can withstand operating voltages of at least 300 volts.

The timed cycle of IC2 is determined by the values of R5 and C4, and will be about 70 seconds if the values shown in the schematic are used. If you desire a longer or shorter listening time when the telephone is answered by The Babysitter, you may increase or decrease the value of either component accordingly. Do not exceed a value of 10 megohms for R5, and be sure to use a high quality low leakage tantalum capacitor for C4.

As with all printed circuits containing integrated circuits, it is recommended that sockets be used for the ICs. This will permit ease of service if any trouble should occur when the unit is first placed in operation, and will facilitate test and checkout of the circuit. Once you have soldered a multipin IC into a printed circuit board it is almost impossible to remove it without destroying the board or IC. Pay careful attention to the location of pin 1 of each integrated circuit as shown in Figs. 18-3 and 18-4. Note that IC3 is placed in the opposite direction as IC1 and IC2. Pin one of each IC is marked at the top of the plastic case by a small dot or numeral 1. Pin 1 as shown in the illustrations is indicated as a small dot next to pin 1.

When mounting the diodes, electrolytic capacitors, and Q1, be very careful to double check the direction in which these components are placed in the board. The circuit will not work if these components are mounted incorrectly. The power-supply bridge circuit has been designed so that all diodes face the same direction. This will help avoid incorrectly placed diodes. If you are using a different diode in the place of CR3, do not inadvertently misplace it in the power supply bridge section.

You may connect the microphone to the circuit using the required length of wire. Be sure to use microphone cable or shielded wire for this connection, otherwise the sensitive microphone amplifier circuit will pick up and amplify the 60-Hz electric field present everywhere in the room. Pay careful attention when connecting the microphone and cable, so that the grounded side of the microphone and cable shield are connected to the grounded, rather than the ungrounded, side of the printed circuit.

After you have completed assembly of the printed circuit board, examine it carefully for short circuits between IC pins, and open circuits along the copper paths. If this is done you should have no trouble when placing the circuit in operation.

CIRCUIT TEST

Before placing The Babysitter in operation use the following procedure to ascertain that the circuit is working properly.

Before applying power to the circuit, remove the three integrated circuits, and fuse the unit with a ½ ampere slow-blow fuse. Insert the power cord into a 120-volt ac receptacle, and measure the voltage across C9 with a dc voltmeter. This voltage should be about 8 volts dc. Connect the negative lead of the voltmeter to any ground point in the circuit, and measure the voltage at pin 14 of IC1, pins 4 and 8 of IC2, and pin 4 of IC3. The voltage measured at these four points should be about + 8 volts dc. If you do not measure the correct voltage as specified above, do not proceed further until the source of trouble has been corrected.

Disconnect the line cord from the power receptacle. Insert the three integrated circuits in their respective sockets, paying careful attention to orient them in the correct direction as shown in Fig. 18-4. IC3 is placed in the opposite direction as IC1 and IC2.

Apply line power to the unit. Connect the negative lead of the voltmeter to any ground point in the circuit and carefully measure the voltage at pin 4 of IC1, without shorting adjacent pins of the IC. This voltage should measure about 8 volts. Measure the voltage at pin 3 of IC1. This voltage should be zero. Take a short piece of wire, and momentarily short pin1 of IC1 to ground. Measure the voltage at pin 4 of IC1. This voltage should now be zero. Measure the voltage at pin 3 of IC2. This voltage should measure about 7 volts, and should hold for about 70 seconds at which time it should return to zero. When this occurs, the voltage at pin 4 of IC1 should return to 8 volts. If the circuit performs as specified, it is ready to be placed in operation.

INSTALLATION AND OPERATION

The Babysitter is connected to the telephone line using the two telephone line terminals as shown in Fig. 18-4 and the schematic diagram. Before making connection to the telephone line you will have to determine the polarity of the line by using a dc voltmeter capable of measuring 50 or more volts dc. When you have identified the positive and negative leads of the telephone line you can properly connect the circuit, observing correct polarity. Be sure to use a standard 4-prong telephone connector or modular plug as required by the FCC Rules and Regulations.

As an initial adjustment, set volume control R15 to midposition. Insert the power cord of The Babysitter into a 120-volt ac power receptacle. Place the microphone at any desired location. The Babysitter is now in operation.

The only way you can check on the operation of the unit is to call your own telephone number from another telephone line. When you do this, The Babysitter will answer your call immediately, and you will be able to hear sounds picked up by the microphone. If you operate a small radio in the same room with the microphone, you will be able to adjust the volume control level as required. This procedure would be greatly simplified if you have someone help you. After The Babysitter goes through its first timed cycle, you can have the other person disconnect the unit from the telephone line, and you can call again with instructions to adjust the volume control, if necessary. Once the unit is properly adjusted, disconnect it from the telephone line until you are ready to use it. Otherwise, anyone who calls you will be greeted by mysterious sounds from the microphone.

Chapter 19

Telephone Computer Memory

A few years from now you probably will be able to obtain, from the telephone company as standard equipment, a "smart" telephone which will memorize dozens of telephone numbers which can be recalled or dialed by the push of a button. Such telephones are available today from independent suppliers who have entered the telephone accessory market in response to the latest FCC rulings on privately owned telephone equipment. These telephones take advantage of the latest advances in LSI (large scale integration) and digital techniques which are being developed by various integrated circuit manufacturers. One of the most recent developments in these repertory dialer chips has been developed by American Microsystems, Inc. You can benefit from these developments right now without investing a great deal of money in a commercially available unit. The Telephone Computer Memory described here (Fig. 19-1) is an easy to build unit which can be constructed at relatively low cost, yet will perform all the repertory dialing functions of a commercial unit.

In order to keep construction costs down, the AMI repertory dialer does not include such features as clocks, timers, calculators, and multidigit readouts. Instead, all of the functions are directly related to storing and dialing telephone numbers, with capability to redial or recall any stored number at will. By using a standard RAM (random access memory) it is possible to store a total of 32 eight-digit numbers, or 16 sixteen-digit numbers. Selection of the memory capability of the unit is provided by a built-in selector switch.

The Telephone Computer Memory makes use of a single-digit

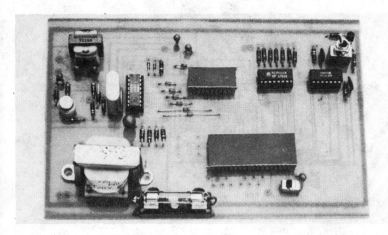

Fig. 19-1. Telephone Computer Memory.

readout, which automatically displays any of the stored numbers at will. The entire number is displayed, one digit at a time, at a rate of ½ second on, ½ second off. All dialing, storing, and retrieving functions are accomplished through a 36-button keyboard (Fig. 19-2). The circuit has been designed so that ordinary single-pole single-throw switches can be used to make up the keyboard. This avoids the problem and expense of obtaining a special keyboard with matrixed circuitry.

The unit includes provision to prevent loss of stored numbers in the event of a power failure. This is accomplished by including a three-cell standby battery which prevents loss of power to the memory chip. During normal operation, power to the unit is supplied by the 120-volt ac power line, and the three-cell emergency power source is kept fully charged. The automatic battery memory preservation feature can be deleted, if desired. In its place is a latching circuit that will light an LED indicator if the voltage to the unit falls below a predetermined value. This feature ensures that at no time will the user try to recall a number from the memory if that memory has been erased by a temporary power failure.

The Telephone Computer Memory provides standard Touch Tone frequencies in response to either the built-in memory, or push-button keyboard operation. The unit is designed as an ancillary interface to the telephone line. This means that it can be connected directly across the telephone line without making any connections to the telephone itself.

Fig. 19-2. Typical 36 key push-button assembly for Telephone Computer Memory.

This feature allows the Telephone Computer Memory to be moved from one telephone to another at any time, if so desired. Since the ringer equivalence of the Telephone Computer Memory is zero, it draws no current from the telephone ringing signal. Thus, you should be able to connect it legally to your telephone line with no additional tariff imposed by the telephone company.

ABOUT THE CIRCUIT

Refer to Fig. 19-3. The Telephone Computer Memory makes use of a CMOS controller chip, IC1, which has been designed for storing, retrieving, redialing, and displaying a maximum of 32 eight-digit (or less) telephone numbers. The storage of these numbers is provided by a 256 by 4 random access memory (RAM), IC4. The dialing or display of these telephone numbers is controlled by a built-in RC oscillator which controls the speed at which the digits are generated. The Telephone Computer Memory makes use of a standard tone generator integrated circuit, IC5, to generate the required signals for Touch Tone operation. The frequency standard for the tone generator is a common 3.58-MHz crystal which is used in color television sets, and is readily available from electronics parts suppliers.

The keypad consists of a bank of 36 single-pole, single-throw switches arranged in a 6 by 6 format. Data entry to IC1 is provided by six column wires and six row wires. The circuit has been designed so that when any one button is pressed the corresponding column wire and row wire for that key are shorted together. Such an arrangement eliminates the need for a special matrixed type keypad which is common to many Touch Tone systems.

For normal dialing functions the digits are entered through the keyboard at any dialing rate. Dial tones are generated and transmitted at the same rate the user enters the digits. The last number dialed is retained in the memory of IC1 and can be redialed by pressing the REDIAL key. As the number is transmitted, the readout displays the stored digits, one at a time.

Storage of the telephone numbers is acommplished by pressing the "*" key. This instructs the controller chip that a new number is to be stored in the RAM. After the digits are entered into the circuit, the "Store" and location button are pressed to instruct the memory to retain the number at the desired location. Any of the stored numbers can be recalled from memory and displayed on the readout, one digit at a time. The rate at which the digits are displayed is controlled by R1, R2, and C1 which control the frequency of the built-in RC oscillator.

In order to display the telephone number one digit at a time, IC1 provides a four-bit bcd (binary coded decimal) number to IC3 which converts the binary information into the proper logic levels to drive the readout and display the desired number. Pin 5 of IC3 is used as an enable lead so that each digit is displayed at the proper time after the data output of IC1 has changed to the desired number. The information fed to pin 5 of IC3 is provided by pins 5 and 6 of IC1, and is processed by two sections of IC2 before it reaches IC3.

Tone generator IC5 is interfaced with both IC1 and IC4, so that it generates the required tones when instructed to do so by either the pressing of a keypad number, or the request to dial a number from the memory of IC4. The tone output of IC5 is coupled directly to the telephone line through isolation transformer T2 and coupling capacitors C4 and C5. This type of arrangement allows the Telephone Computer Memory to be connected directly across the telephone line with no modifications made to the telephone itself. Such a connection is called an ancillary interface to the telephone line.

IC1 contains a power-failure detector which causes a low output voltage to appear at pin 27 if the input power-line voltage falls below a predetermined value. This low-level signal is coupled to two sections of IC2, which are connected in a configuration known as a bistable or latch circuit. Once the voltage at pin 8 of IC2 falls to a logic zero level, the output voltage at pin 10 assumes a logic one level and stays there

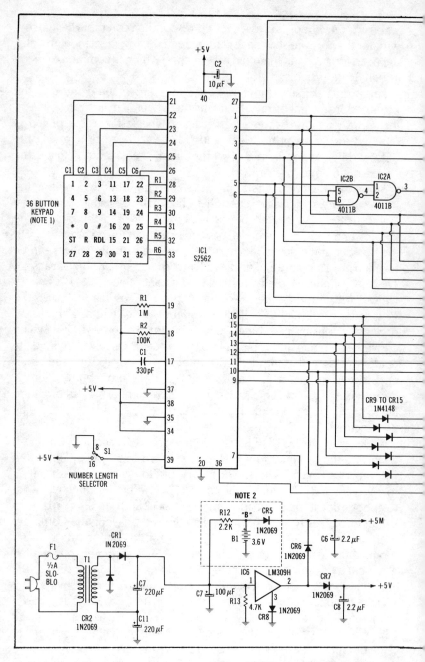

Fig. 19-3. Schematic diagram of the

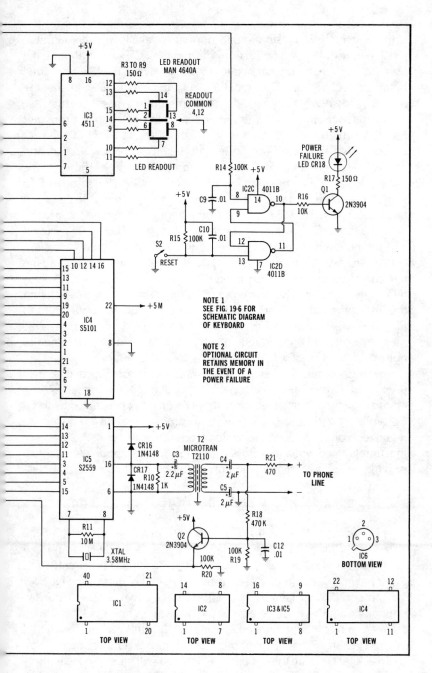

Telephone Computer Memory.

even though pin 8 is restored to its normal voltage level. The output at pin 10 drives Q1 into conduction, illuminating CR18 and alerting the user that a power failure has occurred. A reset switch has been included in the latch circuit of IC2 so that the LED can be extinguished after the memory has been restored by reprogramming the numbers into it.

An optional circuit has been included in the Telephone Computer Memory which will prevent loss of memory in the event of a power failure. This is the circuit composed of B1, a 3.6-volt NiCad battery and its associated components. During normal operation of the unit, the battery is kept fully charged by a trickle current flowing through R12 from the unregulated dc supply voltage. The terminal voltage of the battery is about 3.6 volts, and is insufficient to overcome the reverse bias applied to CR5 by the regulated 5-volt supply. Thus, CR5 is cut off and the battery delivers no current to the unit. When a power failure occurs, the 5-volt regulated supply goes to zero output voltage, and the terminal voltage of the battery is fed to the power input pin of IC4 which retains its memory. CR6 prevents the battery current from flowing into the rest of the circuitry, so that the current from the battery is used to maintain the memory only, and not feed the rest of the unit. This memory protection option requires the use of a RAM chip designed to hold its data with a supply voltage as low as 2 volts.

Power to operate the Telephone Computer Memory is provided by a 6-volt transformer feeding a capacitive input full-wave bridge rectifier circuit. The output of the rectifier, about 8 volts dc, is regulated to about 5.7 volts by IC6. The 0.7-volt voltage drop across CR7 reduces the regulated voltage to 5.0 volts which feeds the entire unit. A portion of the unregulated voltage across C7 is fed to the optional battery to provide a trickle current to keep the battery fully charged at all times.

CONSTRUCTION

The parts list is given in Table 19-1. Most of the circuitry of the Telephone Computer Memory is contained on a single-sided printed circuit board measuring about 4½ inches × 7 inches (11 cm × 18 cm). Refer to Fig. 19-4 for the full-size printed circuit layout and to Fig. 19-5 for the component layout. The printed circuit makes use of narrow copper conductors, since some of the wiring must pass between adjacent pins of the integrated circuits. Best results can be obtained if you use the art work method to produce this printed circuit board. If you use masking tapes on the copper directly, you will have to be extremely careful when etching so that the narrow conductors are not eaten away by the etchant.

Table 19-1. Telephone Computer Memory Parts List

Item	Description
C1	330-pF, disc capacitor
C2	10-µF, 15-volt tantalum capacitor
C3	2.2-µF, 15-volt tantalum capacitor
C4	2.0-µF, 150-volt electrolytic capacitor
C5	2.0-µF, 150-volt electrolytic capacitor
C6	2.2-µF, 15-volt tantalum capacitor
C7	220-µF, 10-volt tantalum capacitor
C8	2.2-µF, 15-volt tantalum capacitor
C9	0.01-µF, 25-volt disc capacitor
C10	0.01-µF, 25-volt disc capacitor
C11	220-µF, 10-volt tantalum capacitor
C12	0.01-µF, 25-volt disc capacitor
CR1 thru CR8	1N2069 silicon diode or equivalent
CR9 thru CR17	1N4148 silicon diode or equivalent
CR18	LED, Xciton XC209R or equivalent
IC1	American Microsystems S2562 repertory dialer
IC2	CD4011B COS/MOS quad NAND gate
IC3	CD4511B bcd-to-7 segment decoder
IC4	American Microsystems S5101 RAM—Note: for optional memory retention use S5101L
IC5	American Microsystems S2559 tone generator
IC6	LM309H National Semiconductor or equivalent
F1	½ amp slow-blow fuse
S1	Spdt toggle switch
S2	Spst spring return toggle switch
B1	3 NiCad "C" cells
R1	1-megohm, ¼-watt, 10% composition resistor
R2	100K, ¼-watt, 10% composition resistor
R3	150-ohm, ¼ watt, 10% composition resistor
R4	150-ohm, ¼ watt, 10% composition resistor
R5	150-ohm, ¼ watt, 10% composition resistor
R6	150-ohm, ¼ watt, 10% composition resistor
R7	150-ohm, ¼ watt, 10% composition resistor
R8	150-ohm, ¼ watt, 10% composition resistor
R9	150-ohm, ¼ watt, 10% composition resistor
R10	1K, ¼-watt, 10% composition resistor
R11	10-megohm, ¼-watt, 10% composition resistor
R12	2.2K-ohm, ¼-watt, 10% composition resistor
R13	4.7K, ¼-wattt, 10% composition resistor
R14	100K, ¼-watt, 10% composition resistor
R15	100K, ¼-watt, 10% composition resistor
R16	10K, ¼-watt, 10% composition resistor
R17	150-ohm, ¼-watt, 10% composition resistor
R18	470K, ¼ watt, 10% composition resistor
R19	100K, ¼ watt, 10% composition resistor
R20	100K, ¼ watt, 10% composition resistor
R21	470-ohm, ¹/₄-watt, 10% composition resistor
Q1, Q2	2N3904 npn silicon transistor

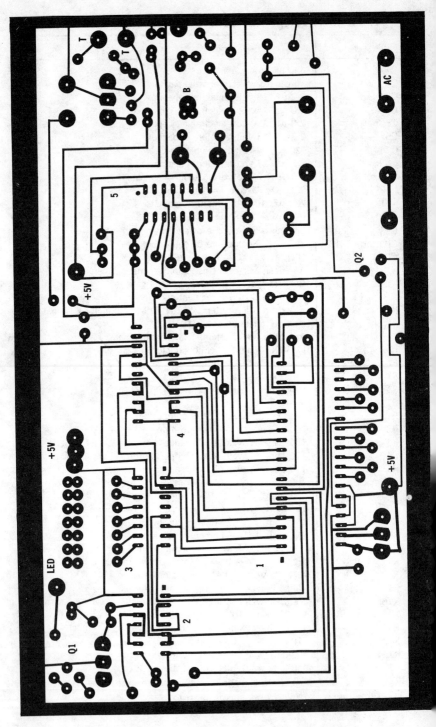

Fig. 19-4. Printed circuit layout shown full size (foil side).

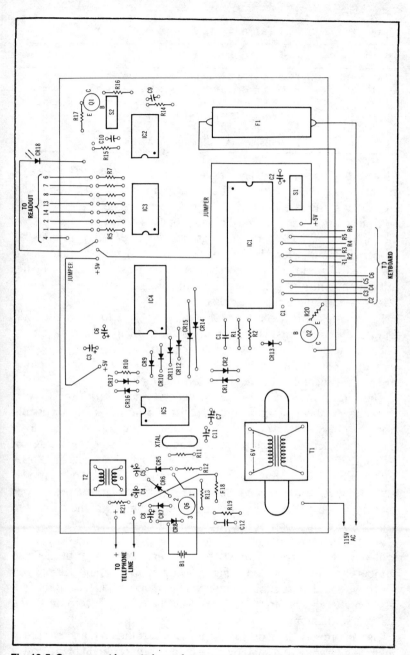

Fig. 19-5. Component layout shown from component side of board, with wiring to external keyboard and readout.

163

Table 19-1. *Continued*

Item	Description
Keyboard	Set of 36 spst, normally open spring return switches
T1	6-volt transformer Radio Shack 273-1384 or equivalent
T2	900 ohm-to-900 ohm telephone coupling transformer, Microtran T2110 or equivalent
Xtal	3.58-MHz color tv crystal
Readout	Monsanto MAN4640A or equivalent

Be sure to include sockets for the integrated circuits. IC1 is not an inexpensive chip, and you will not want to apply power to it until you are sure that the rest of the circuitry has been checked out. The same holds true for IC4 and IC5, which are not as expensive as IC1. Before mounting the IC sockets consult Figs. 19-4 and 19-5 to determine the proper orientation for pin 1 of each chip. Most sockets have pin 1 identified by a small notch or bevel in the plastic. Because of the intricate circuitry, the ICs do not all face the same direction. Double check before soldering the sockets in place.

Follow the component layout shown in Fig. 19-5 for the diodes and electrolytic capacitors. These components are polarized and the circuit will not work if any of them are placed in the board in the wrong direction. The circuit makes use of two types of diodes; place them in the correct locations as shown in the diagram.

The circuit makes use of two switches in addition to the keypad assembly. You will probably not want these controls readily accessible from the outside of the unit, since they are not operating controls. S1 is the number length selector which gives you the choice of storage of either 32 eight-digit numbers, or 16 sixteen-digit numbers. S2 is a reset switch which places the power failure latch circuit in standby mode, and extinguishes the power failure LED. You may place these switches on the printed circuit itself, or place them at the rear of the cabinet in which you plan to house the Telephone Computer Memory.

After all components are mounted on the printed circuit board, you will have to wire two jumpers for the regulated 5-volt bus. These jumpers are clearly marked in Fig. 19-5. In addition to this wiring, you will have a multiwire cable connecting the keyboard, LED, and readout assembly to the printed circuit. Follow the wiring according to the schematic diagram.

Use an ordinary light duty line cord to power the unit, and be sure to include a ½-ampere slow-blow fuse in the primary circuit of the power transformer. When making the connection to the telephone line use either a standard 4-prong telephone plug, or the new style modular connector. This is a requirement by the FCC and should be followed without fail. Before wiring the telephone connector plug, check the polarity of your telephone line with a dc voltmeter capable of measuring at least 50 volts dc. Then wire the plug so that C4 and C5 are connected to the telephone line with the correct polarity. Accidental reversal of the voltage across the two capacitors will cause them to draw excessive current, leading to failure of these parts.

If you plan to use the optional battery standby circuit to protect the memory in the event of a power failure, secure the cells inside the cabinet. Be sure to pay strict attention to the polarity of the cells. It is best if you work with cells which are not charged, since an accidental short circuit across one or more cells can cause many amperes of current flow. Such current can melt wiring insulation and board conductors. Once the circuit is placed in operation the trickle charge provided by the power supply will keep the cells fully charged. Note: When including the battery-operated memory protection circuit, you must use

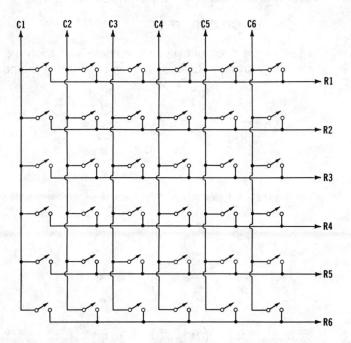

Fig. 19-6. Schematic diagram of keyboard. Switches are spst, normally open, spring return. See Fig. 19-7 for switch identification.

a special low-voltage version integrated circuit for IC4. Refer to the parts list for the proper part number.

The mounting of the keyboard switches and readout is left up to the builder, since it will depend upon the style of components you plan to use. It is best if you use a standard integrated circuit socket for the readout. This will permit easy replacement should it ever be necessary. Follow the schematic diagram for the keyboard assembly, Fig. 19-6. The readout wiring is shown on the main schematic diagram for the unit.

After you have completed wiring the entire unit, it should be carefully checked against the schematic diagram for wiring errors. This procedure will help avoid problems when the unit is placed in operation, and may prevent damage to one or more of the components. Do not place any of the integrated circuits into the sockets until instructed to do so in the checkout procedure.

The keyboard layout is shown in Fig. 19-7. Note that "0" and "#" push buttons are also used for locations 10 and 12. A typical printed circuit layout for a keyboard assembly is shown in Fig. 19-8.

CIRCUIT TEST

The first part of the checkout procedure consists of applying line power to the unit with only IC6 installed, and measuring the power

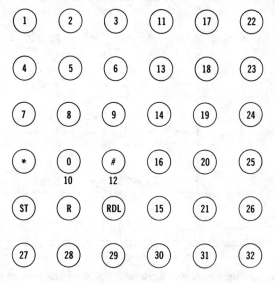

Fig. 19-7 Keyboard layout.

supply voltages at each IC socket according to the pin numbers specified in the schematic. Pin 2 of IC6 is the output of the regulator and should measure about + 5.7 volts with respect to circuit ground. If this voltage measures normal, then check the power input pins of IC1 IC2, IC3, IC4, and IC5 for a voltage of about + 5 volts. Measure the voltage at each grounded terminal of those ICs as shown in the schematic diagram. The voltage reading at these points should be zero. Do not proceed further until you are satisfied that the voltage measurements specified above are all correct.

Disconnect the line power to the unit before installing the integrated circuits in their respective sockets. This cannot be overemphasized. When installing the integrated circuits be very careful not to bend the pins, and check to be sure that no pin has slipped underneath the plastic case instead of into the socket. After all integrated circuits have been installed in the circuit board, recheck the position of each IC to ascertain that it is inserted into the socket in the correct direction. Refer to Fig. 19-5 for the correct orientation of these parts. After all ICs have been installed, you are ready to proceed with the checkout procedure, which is actually putting the circuit through its various operating modes. This test should be performed with the Telephone Computer Memory connected to the telephone line, so that the telephone handset can be used to feed back tone information to you. This will provide information that the circuit is working properly. Use the following procedure:

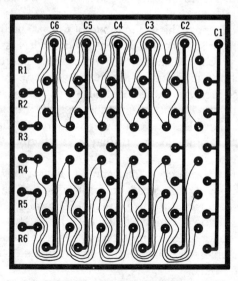

Fig. 19-8. Typical printed circuit layout for keyboard assembly shown half-size.

1. *Normal Dialing*—Enter the desired telephone number, after lifting the telephone handset off the hook, by pressing the appropriate buttons of the keyboard. As the numbers are entered, you should hear the generated tones in the handset, and the display should read out the correct digits.

2. *Redialing*—To check the redialing capability of the unit, hang the telephone up momentarily, lift the handset, and press the "redial" key (RDL). The number that was previously dialed will now be transmitted as the display exhibits each digit.

3. *Telephone Number Storage*—This is performed with the telephone "on the hook." Push the "*" key. Press the desired keys for the number to be stored. Do not exceed 8 digits if the unit is set for a total of 32 telephone number storage. Push the "ST" key. Push the key (1 to 32) corresponding to the desired location of the stored telephone numbers until the memory is filled to capacity.

4. *Displaying the Last Dialed Telephone Number*—Push the "RDL" key. The last dialed number will be displayed, one digit at a time.

5. *Displaying a Number Stored in the Memory*—This is performed with the telephone "on the hook." Press the "R" key. Press the key corresponding to the desired telephone number location. The stored telephone number will be displayed, one digit at a time.

6. *Repertory Dialing*—This is the mode of usage which will be most used with the Telephone Computer Memory. It is assumed that the user has prepared a chart which indicates the address location of all telephone numbers stored in the memory. After lifting the telephone receiver off the hook, push the "*" key. Then push the key corresponding to the location of the desired telephone number. The Telephone Computer Memory will dial the number out, with a simultaneous display of each digit as it is being transmitted.

7. *Power Failure Detection*—You can simulate a power failure by simply removing the power cord from the ac receptacle. When the power is restored, the power failure LED should be lighted. Press the reset button to extinguish the LED. Note: This test will cause your Telephone Computer Memory to lose all stored telephone numbers unless you have equipped it with the optional memory protection battery.